Food, Politics, and Society

The publisher and the University of California Press Foundation gratefully acknowledge the generous support of the Ahmanson Foundation Endowment Fund in Humanities.

Food, Politics, and Society

Social Theory and the Modern Food System

Alejandro Colás, Jason Edwards, Jane Levi, and Sami Zubaida

UNIVERSITY OF CALIFORNIA PRESS

University of California Press, one of the most
distinguished university presses in the United States,
enriches lives around the world by advancing scholarship
in the humanities, social sciences, and natural sciences. Its
activities are supported by the UC Press Foundation and
by philanthropic contributions from individuals and
institutions. For more information, visit www.ucpress.edu.

University of California Press
Oakland, California

Library of Congress Cataloging-in-Publication Data

Names: Colás, Alejandro, author. | Edwards, Jason, 1971
    October 3- author. | Levi, Jane, author. | Zubaida,
    Sami, 1937- author.
Title: Food, politics, and society : social theory and the
    modern food system / Alejandro Colás, Jason
    Edwards, Jane Levi, and Sami Zubaida.
Description: Oakland, California : University of
    California Press, [2018] | Includes bibliographical
    references and index. |
Identifiers: LCCN 2018009662 (print) | LCCN 2018012663
    (ebook) | ISBN 9780520965522 (ebook) |
    ISBN 9780520291942 (cloth : alk. paper) |
    ISBN 9780520291959 (pbk. : alk. paper)
Subjects: LCSH: Food—Social aspects—History. |
    Food—Political aspects—History.
Classification: LCC GT2850 (ebook) | LCC GT2850 .C57
    2018 (print) | DDC 641.3—dc23
LC record available at https://lccn.loc.gov/2018009662

Manufactured in the United States of America

27 26 25 24 23 22 21 20 19 18
10 9 8 7 6 5 4 3 2 1

Contents

Preface and Acknowledgments

The origins of this book lie in an undergraduate course entitled "Food, Politics, and Society" which all four of us have taught together at Birkbeck College, University of London, since 2011. Sami Zubaida and Jane Levi have a long-standing involvement in the study of food and society, while Jason Edwards and Alex Colás arrived more recently to this field of study through their research on politics and international relations, respectively. A combination of commensality and a shared interest in social theory (broadly conceived) eventually led us to design a syllabus encouraging students to think about politics and society through the everyday practices of eating and drinking. In teaching this course from the perspective of political sociology prominent in our department, it became apparent to us that modern social theory and the contemporary food system were more closely intertwined than is often acknowledged. The present book is therefore both an invitation to consider more systematically the centrality of food and drink in the development of modern social theory, as well as a historical sociology of the modern food system through the examination of a dozen key concepts.

We have benefitted enormously from colleagues at the University of London and elsewhere in thinking about and writing this volume. The Birkbeck Institute for Social Research has sponsored our College Food Group, which for almost a decade has hosted diverse public events and research seminars with contributions from numerous speakers that have informed much of the book's content. Our neighbors in the School of Oriental and African Studies' Food Studies Centre have generously promoted our gatherings, as has the British Sociological Association's Food Study Group. We are grateful to these colleagues, academic networks, and institutions for, however indirectly, supporting our intellectual endeavors. Four anonymous reviewers read a first draft of the manuscript, offering many helpful insights and suggestions, most of which we have incorporated into the final version. Our thanks go to them and also to Kate Marshall and Bradley Depew at the University of California Press for seeing this project through to fruition. Needless to say, we alone are responsible for any remaining omissions or errors. Our final expression of gratitude goes to our Food, Politics, and Society students, since they offered us an audience in testing out our ideas and have thus in many unexpected ways participated in the production of this text.

Alex Colás, Jason Edwards,
Jane Levi, and Sami Zubaida
London, January 2018

CHAPTER I

Introduction

Food, Drink, and Modern Social Theory

In his famous article on the sociology of the meal, Georg Simmel (1858–1918) observes that "of all the things that people have in common, the most common is that they must eat and drink."[1] This truism, however, comes with a paradox insofar as the "exclusive selfishness of eating" (that is, the necessarily individual act of ingesting food and drink) mostly overlaps with the "frequency of being together" (eating and drinking in society). "Because this primitive physiological fact is an absolutely general human one," Simmel continues, "it does indeed become the substance of common actions."[2]

The present book is about the interaction between that exclusive selfishness of eating and drinking and the common actions of society this basic physiological fact engenders. We show how modern social theory can illuminate and explain many of the processes and institutions that have resulted from people eating and drinking in society, and how, in turn, much modern social theory has been informed—sometimes directly, generally circuitously—by specific patterns of food production, preparation, and consumption. Each chapter in the book focuses on a set of key concepts in modern social theory that shed light on the structures and dynamics of the relationship between food, politics, and society today and in the past. If we take, for instance, the photograph by Josep María Sagarra that graces this book's cover, it shows men and women consuming food and drink in close company. We know from the title of the image that it was shot at a fundraising party for a Red Cross Hospital in

Barcelona in 1932. We might also surmise from the guests' formal and bejeweled attire, the linen and silverware that adorn the table, and the large mirrors and chandeliers that furnish the room that this was an occasion open only to the city's high society. It was likely the type of reception where people come and eat and drink in common, but not one where common people come to eat and drink.

There are, therefore, immediate reflections to be made on class and gender as they're represented in this picture. And if we take a further step back to think of how the food and drink got to that table and how the room was set and subsequently cleared up, all sorts of other social relations involving the production, processing, preparation, and serving of food and drink, as well as their consumption and disposal, come into view.[3] What relations of production facilitate the common acts of consumption at the fundraising party? Why is the food and drink taken standing, and with cutlery seemingly lying idle? Why offer such a spread at a charity event for a humanitarian organization? Who selected the wine, and who washed the dishes? These sorts of banal questions shape the chapters that follow because they speak to some of the grand themes of modern social theory since its inception in the late seventeenth century. The separation between the private and public spheres of social life; competing conceptions of identity, belonging, and community; diverse notions of distinction, civility, and taboo all permeate the common act of eating and drinking, and all have also been central to the development of modern social and political theory. Moreover, the defining socioeconomic and political transformations of the modern period—urbanization, industrialization, rationalization, commercialization, democratization—have clearly impacted the production, preparation, and consumption of food and drink as much as they've articulated the principal concerns of modern social theory. In fact, as we'll try to demonstrate, food and drink has been a focal point of many more classic studies in social and political theory than is often acknowledged—from Habermas's political writings on the public sphere of the coffeehouse to Bourdieu's sociological reflections on gastronomic "distinction" and "habitus" and from Mary Douglas's and Claude Lévi-Strauss's anthropological musings on food prohibitions and cuisine to Amartya Sen's political economy of famine. It is, of course, telling that these represent a selection of late twentieth-century authors expressly concerned with food and drink and not the earlier Western canon of Durkheim, Marx, and Weber, among others.[4] But, as the rest of the book endeavors to show, many of the ideas of these great luminaries have been adopted and extended over the past few decades to build up a formidable corpus of food-related social and political theory that cuts across old and new disciplines like sociology, cultural studies, environmental history, global political economy, gender studies, anthropology, and political philosophy.[5] It is our ambition in the pages that follow to convey some of the richness emerging from this combination of social and political theory with the study of food and drink.

The book thus sets itself the tall order of making huge comparisons between big structures and large processes attached to the modern food system.[6] We adopt an approach broadly identified by Stephen Mennell, Anne Murcott, and Anneke van Otterloo as "developmentalist," in that it emphasizes the changing nature of the relationship between food, politics, and society across time and place, although there are also some "materialist" strains present in our understanding of the socio-ecological determinants of such interactions.[7] We aim to let our theoretical insights emerge from the historical-sociological narrative, rather than impose some tight, parsimonious theoretical framework on the wide-ranging experiences conveyed below. There are, however, a number of conceptual threads running across the following chapters that require some clarification and explanation. The rest of this introduction seeks to do this, first, by defining some of our core terms and showing how they relate to the modern food system and, second, by outlining how the various chapters apply diverse social theories and their associated categories in explaining the dynamics of the modern food system.

DEFINITIONS: THEORY, MODERNITY, SOCIETY

Modern social theory crystallized as a distinctive way of thinking about human affairs in the course of the 1700s, in response to what Bruce Mazlish called the "breakdown of connections."[8] Whereas in most parts of the world, and in Europe in particular, human societies had until then been organized around political units that connected people and nature through a fairly static hierarchical order legitimized and enforced by religion and otherworldly cosmologies, the arrival of modernity was marked by the unshackling of multiple socioeconomic, political, and ideological fetters in the form of inherited privileges, codified rank, clerical rule or restrictions on trade and economic activity. "A great tectonic shift seemed to be taking place," Mazlish suggests, that "proclaimed itself in an omnipresent, even compulsive concern with the snapping of ties, the unchaining of all established verities and social arrangements."[9]

The origins and periodization of this radical change—and its main drivers—are, of course, still the subject of heated debate in the social sciences. For the conceptual historian Reinhart Koselleck, the hundred years from 1750 to 1850 represented a "threshold period" *(Sattelzeit)* in European history in which, spurred on by the French Revolution and the Enlightenment (themselves conceptual progeny of the *Sattelzeit*), ancient categories like "democracy," "nation," "civil society," or "culture" were reappropriated and transformed into basic concepts—terms that are indispensable when understanding the socioeconomic and political structures and processes of modernity, and also, without which, we moderns cannot make sense of our own time.[10] Other historians of ideas, such as J. G. A. Pocock or Quentin Skinner, have underlined the rise of secular (i.e., time-bound, this-worldly) understandings of politics and society during the European Renaissance and Reformation, which, in turn, generated the modern institution of the sovereign territorial state, whose absolute authority increasingly trumped that of seigneurial or ecclesiastical jurisdictions.[11] For their part, thinkers of the Scottish Enlightenment claimed it was the complex division of labor, commodity exchange, and widespread extension of private property rights that delivered a modern commercial or civil society where, as Marx and Engels would have it, "all that is solid melts into air, all that is holy is profaned, and man *[sic]* is at last compelled to face with sober senses his real conditions of life, and his relations with his kind."[12]

Without imposing some false uniformity across all of the chapters, we adopt in this book many of these claims made for modern social theory as a body of thought that both *emerges from* and *reflects upon* the systematic breakdown of connections that began with the long sixteenth century (1450–1650) and arguably continues into the present day. Here, "theory" simply involves the process of critical reflection or contemplation on the causes and consequences of human agency—both individual and collective—in the development of enduring socioeconomic and political phenomena. In other words, producing concepts that account for what Émile Durkheim (1858–1917) called "social facts."[13] It includes the analysis of politics too, understood as the processes and institutions of government that have emerged from living together in a *polis*—a city or spatially delimited community that abides by given rules, procedures, and practices of power. We therefore use social and political theory interchangeably, only singling one out from the other for purposes of emphasizing the informal, everyday dynamics of the former and the more formal, institutional character of the latter.

In both cases, however, there is a recognition that "theory" and "practice" are deeply intertwined (concepts always operate within a concrete social context), and that this relationship changes through place and time (ideas, practices and their contexts vary geographically and can be transformed historically). We are, moreover, mainly engaged in what Nicos Mouzelis once described as "sociological theory": the application of "conceptual tools for looking at social phenomena in such a way that interesting questions are generated and methodologically proper linkages established between different levels of analysis."[14] It is not the task of this book to present an entirely new, substantive theory, but rather to put to work existing conceptual frameworks and paradigms in the explanation of the interaction between food, politics, and society.

This all said, "modernity" serves in this book to identify a distinctive historical period and condition, ranging from the long sixteenth century to the present, where certain isms and izations (including capitalism, nationalism, socialism, racism, feminism, individualism, secularization, industrialization, rationalization, and commodification) have become the dominant expressions of human agency. The invocation here of modernity should not be confused with the resuscitation or endorsement of modernization theory, understood as a linear sequence of stages through which all societies must pass through or "skip over." In what follows, we think of modernity as an epoch and condition that not only unfolded in all kinds of uneven and protracted ways across different times and places but also has arguably intensified and combined distinct modern and traditional temporalities or worldviews in, for instance, the recharging of ethnic or religious identities in contemporary food cultures or the unequal globalization of primary food commodities. More specifically, our study addresses three distinctive yet interconnected phenomena that have characterized modern history: transformation, stratification, and globalization.

One of the characteristics of modernity is the self-consciousness of its own temporality, however contrived. Be it the idea of the Renaissance, the Reformation, or the Enlightenment, the historical semantics of modernity imply a radical break with the past. The very notion, for instance, of a Neolithic revolution (first approached in the next chapter) as the birth of agriculture is a modern construct, the product of an evolutionary and secularized understanding of social development that organizes human history along different stages in our collective relationship with nature. Similarly, the idea of "the self"—the individual subject able to make conscious choices and shape his or her own future through independent agency—is a modern creation (as chapters 7 and 11 indicate). Manifest in art, literature, and philosophy, the modern subject also finds intense expression as a customer through our food choices (including, quintessentially, the restaurant menu), as a target of marketing, and in the connections between diet, health, and our bodies. We are therefore especially attuned in this book to the notion of modernity as an eminently revolutionary period, where all sorts of identities, customs, institutions, techniques, and ideas are constantly reformed and transformed.

This is why the Industrial Revolution appears in so many of the chapters below. In chapter 5, we show how the Industrial Revolution forever changed our diets, eating and drinking habits, as well the food system's modes of production and consumption. But it also transformed the physical landscapes and the built environment, reshaping notions of private and public in cities, as well as the relationship between town and country. Moreover, the extensive urbanization that followed created the spaces of consumption where many new political ideologies, commercial enterprises, cultural entities, and social movements were forged. We suggest, somewhat counterintuitively, that the transformation of the early-modern British food system through changes introduced by agrarian capitalism long predate, and in many respects, instigated the Industrial Revolution of the late eighteenth century. Thus, rather than technological innovations of the Industrial Revolution radically transforming agriculture, it was changes in British agriculture from the seventeenth century that paved the way for the subsequent Industrial Revolution. Whatever the causal chain, the Industrial Revolution increased average calorific intake across most human societies, raising life expectancy and thereby contributing to our exponential population growth. It also improved average land yields and agricultural productivity through mechanization, the use of synthetic fertilizers, and artificial irrigation. The "second wave" of the Industrial Revolution introduced words like *pasteurized, refrigerated, canned,* and *tinned* into our gastronomic vocabulary, as well as revolutionizing both household and retail cooking and cleaning through the mass extension of gas and electric lighting, cookers and ovens, food processors, toasters, washing machines, and internal plumbing.

It is important to note that Mazlish refers to both the *making* and *breaking* of connections as the midwife of modern sociology. The deep ruptures that accompanied the birth and development of modernity launched what political economist Joseph Schumpeter (1883–1950) called the "creative destruction" of industrial capitalism, which allows all manner of preexisting social forms—caste, patriarchy, ethnicity, religion, and empires—to be recast and reinvented in the construction of modern patterns of production, processing, and consumption. Hence the socioeconomic and political transformations occasioned by the French and Industrial Revolutions also brought in their wake new expressions of social stratification. The rise of an urban proletariat and its dependents is an obvious example of this. But so are the attendant rearticulations of gender relations, particularly in the household, as women acquired the double burden of salaried work outside the home only to continue their working day as carers and homemakers in the domestic sphere. Social stratification through the production and consumption of food and drink obviously predates modernity, but the modern period witnessed a distinctive reformulation of rank, status, and distinction through what social theorist Norbert Elias (1897–1990) called the "civilizing process."[15] As we indicate in chapter 9, during the long sixteenth century, the European court became a site for the development of table manners—including the protocols on use of forks, knives, and serving devices—as a social mechanism for reinforcing and reproducing the elite status of courtly aristocracy. By the end of the nineteenth century, this attention to the social organization, preparation, and presentation of courtly cuisine was converted by famous chefs, restaurateurs, and hoteliers like Carême, Escoffier, and Ritz into the canon of French haute cuisine whose innovations, we indicate, involved a move away from heavy sauces and toward lighter ones as a marker of the transition from the past to a more delicate, "modern" cuisine. The period also witnessed the shift from the French to the Russian style of serving, which required the service of individual dishes in sequence, signaling a wider trend to greater simplicity and delicacy in food preparation and presentation, which is represented in much fine dining to this day.

The mass migrations (both within and across borders, voluntary and forced) facilitated by capitalist industrialization also reconfigured racial hierarchies and ethnic segmentation within cities and in rural areas. The modern food system has plainly been affected in various ways by these changing structures of class, gender, and ethnic integration, gradation and segregation. In chapter 8—on national, regional, and ethnic gastronomy—for example, we highlight the role of urbanization in the codification of national cuisines through the concentration and admixture of otherwise dispersed and highly regionalized repertoires. State formation was, however, accompanied by the reinvention of class and regional hierarchies as part of a process of national standardization well into the contemporary period. With reference to the recent Turkish experience, that chapter describes how, during the 1980s in particular, the spicier, stronger flavors of Anatolian food came to Istanbul and other major cities along with Anatolian migrations, in the form of kebab grills, known as *Gaziantep,* provoking condescension and disdain from the Istanbul bourgeoisie.

It would be impossible to fully understand the modern experiences of transformation and stratification just alluded to without also referring to a third phenomenon—globalization. This admittedly slippery term acts as shorthand for the wider process of worldwide traffic in goods, peoples, and ideas inaugurated by the European conquest of the Americas. Once again, there is obviously no question such socioeconomic and cultural transfer was occurring long before the advent of modernity (most notably subsequent to the Agricultural Revolution discussed in chapter 2), but the uniquely global dimensions of what environmental historian Alfred W. Crosby Jr. called the "Columbian exchange," allied with the birth of a world market it occasioned, gives the 1492 turning-point a distinctively epochal quality.[16] If our book's focus relies disproportionately on European theories and illustrations, this is not because we wish to endorse some spurious Eurocentric superiority or exceptionalism, but simply because we are keen to underline the sharp structural inequalities within and between states and regions that has resulted from the European colonial expansion since the end of the fifteenth century. The Columbian exchange, we insist in chapter 3, was deeply unequal and uneven— it simultaneously integrated the world into a global economy and fragmented humanity and nature along new political, geographical, and cultural hierarchies. The Columbian exchange was not just about the cross-Atlantic transfer of corn, beans, and squashes in one direction and livestock, sugar, and wheat in the other—it entailed colonial conquest, with all the subjection, oppression, and despoliation this implies. We are therefore alert, throughout this book, to the relations of exploitation and domination that underlie the seemingly innocent use of terms like "fusion cooking" or "creole cuisine," enriching as these often are. Indeed, chapter 10 in particular—on the political economy of the global food system— highlights the continuities in the geographical unevenness and the socioeconomic inequality of the various modern "food regimes." Furthermore, as in other areas of social life, the modern food system has found diverse expressions across different parts of the world—it has been modified, adapted, and challenged by local social forces and cultural traditions. Yet, in line with the dialectic of creative destruction we are also adamant here that, like culture or identity, modernity is never static or one-sided (once again, it is not a series of predetermined stages), but rather it constantly revolutionizes social relations in both time and place. With specific regard to food and drink, this has been especially noticeable in the changing economic and cultural geography of food production and consumption, as many erstwhile colonial societies (think of South Africa, Vietnam, Brazil, or Ireland) have become major players in regional and global food and drink sectors.

These, then, are some of the common denominators that bind together the otherwise diverse themes covered in the book: a focus on the acceleration and intensification of social life during the loosely defined period of modernity, an emphasis upon the new or reconfigured social cleavages this epoch has produced, and a resolutely globalist approach to social change and stratification that constantly probes the transnational and international dimensions of the modern food system. Here, "society" and the "social" refers to a "reciprocity of strangers" that gradually but irrevocably replaced the prevailing hierarchies of community in the organization of human life, and "modern social theory" to the conceptual explanations for the structured processes that characterize this shift. Modernity is understood as both a historical era with a relatively elastic periodization (some argue it began only with industrialization; others, that it ended in the 1970s, giving way to postmodernity), and a specific social condition marked by what Max Weber called the "disenchantment" of the world. The modern food system—a globally integrated market in the production, processing, distribution, and consumption food and drink—is one outcome of our historical period, which, as the next section of this introduction suggests, can be fruitfully analyzed with reference to some key concepts in modern social theory.

SOCIAL THEORY AND THE MODERN FOOD SYSTEM

At its best, social theory renders intelligible the otherwise unfeasibly large number of discreetly individual actions that form society and that cannot be merely described empirically. Like all theory, social theory deals in categories that abstract out the main features of specific phenomena in order to provide some analytical coherence and explanatory purchase on myriad human interactions. This explanatory labor has been undertaken within specific academic disciplines—sociology, anthropology, history, psychology, political science, geography—which have developed their own problematics, debates, research methods, and seminal texts, each reflecting different analytical registers and theoretical preoccupations. Moreover, successive waves of ideological tendencies and methodological paradigms—functionalism, Marxism, feminism, constructivism, structuralism, and postmodernism, among others—have made their way into social theory over the years, including in the study of food and drink. We don't explicitly take sides in those debates or offer any taxonomy of the various social-theoretical approaches to food and drink. Instead, we focus in this book on three pairs of categories that are no less real for being abstractions and that, from our perspective, capture both the macro-sociological spatio-temporal dynamics of modernity and its more intimate and micro-sociological expressions in everyday life. They offer an opportunity to illustrate how many of the categories issuing from modern social theory can explain the modern food system and how, in turn, food and drink have shaped some of the chief concerns of modern social theory. These binaries are the public and private, nature and society, and the self and other.

In February 1960, four black students from the local agricultural university sat at the lunch counter of a Woolworth's store in Greensboro, North Carolina, and ordered a cup of coffee. They were denied service because that lunch counter was, in that state, reserved for whites only. What subsequently became—together with various other sit-ins and occupations of public space—a signal moment in the American civil rights movement clearly tells a political story about segregation, protest, and the struggle for racial justice and equality in the United States. However, it also raises important issues about the separation between the private and the public and the role of food and drink in defining this social distinction. The dynamics of violence and resistance represented opposite, in the iconic photograph by Fred Blackwell, taken in the course of another civil rights sit-in, this time in Jackson, Mississippi, in May 1963, speaks volumes about the complex relationship between individual act of ingesting food and drink and the "substance of common actions" Simmel spoke of.

Were it not for the color of their skin, the "Greensboro Four" and their counterparts in the other sit-in campaigns would have been considered private customers—that is, "sovereign consumers"—entitled to do with their own money as they pleased, engaging in the commonplace market transaction of buying a hot drink at the Woolworth's lunch counter. (Indeed, the Jackson sit-in formed part of a wider boycott of segregated Capitol Street stores, where protestors presented themselves as regular customers, merely demanding "service on a first-come, first-serve basis for all customers—blacks as well as whites").[17] Yet because

FIGURE 1. 1963 Jackson, MS, Woolworth Lunch Counter Sit-In. Photographed by Fred Blackwell.

these private exchanges took place in a public setting, in southern U.S. states where racial discrimination was authorized by law, the simple act of black people ordering coffee or of blacks and whites merely sitting next to each other as equals at a segregated counter immediately acquired a wider social and political significance. It was registered as an individual act of defiance that soon mobilized collective protest, publicized an everyday experience in the lives of African Americans that many of their fellow citizens might otherwise have been oblivious to, and, most obviously, challenged American notions of freedom and equality for all.

Powerful as these local acts of resistance were, they formed part of a longer history in the social and political reconfiguration of public spaces of food and drink consumption, where social action and personal conduct are differentiated from that obtaining in the private sphere. The German critical theorist Jürgen Habermas (1929–) identifies the coffeehouse of early-modern London as one of several social sites responsible for the "structural transformation" of the public sphere during the seventeenth and eighteenth centuries. According to Habermas, these places became more than plain drinking establishments, as they also nurtured distinctively modern, bourgeois forms of debate and communication that, in their emphasis on reasoned argumentation and informed conversation among equals, became an essential plank of democratic deliberation in subsequent centuries. Critics of Habermas have noted how women and artisans, among others, would generally have been excluded from the bourgeois public sphere, thereby limiting the democratic potential of this domain.[18] The essential role of the coffeehouse as a venue of commercial exchange also suggests that the free and egalitarian modes of communicative action Habermas champions were severely compromised by an instrumental rationality dominated by profit and calculation. The sit-ins staged in Greensboro and Jackson thus remind us about the politically contested nature of the public sphere as conceived by Habermas—it can foster both an egalitarian reciprocity of strangers associated to "civil society" and the exclusionary, secretive, or elitist expressions of public sociability, such as that of the whites-only lunch counter at Woolworth's. Yet, as we discuss in chapter 6, none of this diminishes the profoundly political character of coffeehouses, taverns, pubs, and restaurants, notwithstanding their differences. Food and drink consumption in these locales is politically determined in much the same way as the public-private distinction. Drawing on Richard Sennett's work, we argue against theorists like Hannah Arendt (1906–1975) who claim that the "common act of eating" is about (consumer) sameness, not (political) equality. We instead insist that the material spaces and practices of consumption in coffeehouses, alehouses, taverns, or, indeed, lunch counters are not just incidental to political debate, organization, and contestation, but a condition of them.

This is an important insight for our book's premise, since it suggests that food and drink are not merely contingent to Habermas's idea of a bourgeois public sphere, but are a fundamental, substantive component of his account. In focusing on the coffeehouse as a fulcrum of distinctive forms of communicative action, Habermas is also necessarily incorporating, however inadvertently, some of the material properties of food and drink (coffee, in this instance) into his political theory. Coffee was imported into England as part of an increasingly complex international trading network (which also included tea, tobacco, and sugar as other social stimulants); it was, like other imports, taxed by the fledgling state; and its consumption, unlike other tradable commodities, involved a performance of sharing that "primitive physiological act" of ingestion in the company of others. Thus, the truly cosmopolitan character of the coffeehouse, its encouragement of peculiarly modern forms of public sociability, and its place within a wider political economy of profit, trade, taxation, and state regulation (all key elements of Habermas's bourgeois public sphere) are in large measure conditioned by the fact that coffee was a foreign drink. This is plainly not to say that London developed a public sphere because of coffee, but it is to suggest that the attributes of coffee (and, indeed, tea, tobacco, and sugar) shaped London's public sphere, as analyzed by Habermas.

Thinking about the public-private distinction through food and drink also directs us to questions of state authority, market power, and household consumption. Anecdotally, the political vocabulary of many states can be traced to foodstuffs. The words *salary* and *dole* derive from Latin words that, in ancient Rome, denoted, respectively, a soldier's payment (partly in salt) and grain handouts, while the Arabic term *makhzen,* which in Morocco is synonymous with the state, means "granary" or "warehouse"—both pointing to the historic role of food storage as a source of political power. Similarly, the Chinese character for a grain bushel, *shih,* was used by the Han Empire to rank different bureaucratic offices, reflecting the agrarian foundations of this redistributive polity. In chapter 9, "Distinction," we chart the mutual reinforcement in early-modern Europe between state centralization, new norms of civility, and the replacement of a feudal by a courtly aristocracy. Here food played a significant role in bolstering the public authority of the state, not merely through taxation of comestibles (such as the infamous *gabelle* salt levy), but also in the exhibition of courtly patronage and munificence through state banquets and dinners that, though exclusive in their guest lists, had a pronounced performative dimension aimed at a wider public. (There are arguably remnants of this courtly display of food consumption in the fundraising reception captured in the cover image of this book). We contrast this spectacular exhibition of luxury and grandeur with the markedly private, low-key, and secluded dining of the Ottoman court, which remained repetitive and confined, thus illustrating the different geographical expression of modernity. The civilizing process associated to absolutist Europe (the seventeenth- and eighteenth-century control of the state through dynastic power) did reinforce public institutions in various guises, but it was also premised on the personalized rule of the monarch through private patronage. In its combination of public authority with private gain, absolutism produced what Max Weber called a "patrimonial state," and this was captured in the organization of courtly cuisine discussed in that chapter.

On some liberal interpretations, the absolutist conflation of public and private power was undermined by democratic struggles for the separation between these two spheres from the nineteenth century onward, so that individuals' "negative freedoms" over their personal beliefs, identity, and property came to be increasingly distinguished from the "positive liberties" of collective welfare, education, and employment to be secured by the state. The public life of salaried work, political activity, and commercial exchange was thus ideal—typically contrasted to the affective realm of the household, where care, reproduction, and intimacy characterized human relations. The pub, tavern, or restaurant was deemed an arena of the male breadwinner; the home a sphere reserved for the housewife's task of raising and nurturing the nuclear family. Patterns of private household consumption (including the number, content, and timing of daily meals) have thus often been linked to the structures of public life (relating to class, region, and religion): male English workers ate their "(high) tea" after work and before going to the pub, while their bosses would have "dinner" at home or at their private club. Jewish families in postwar Baghdad might be invited to a non-Jewish home to partake in the typical national specialty, *masgouf* (prepared with *shabbout* fish or varieties of carp), previously cooked in the neighborhood baker's oven.[19]

Feminist and other radical critics of the public-private distinction have long argued that the personal is political and that housework plays a critical public function in the reproduction of modern societies. They have, moreover, shown how the stereotypical gendered division of labor underpinning the public-private divide (insofar as it permanently existed) has been challenged by the feminization of the workforce and, indeed, the changing notions of what constitutes gender or a family. With the proliferation of takeout and home-delivery dining, "eating out" is no longer distinguishable from "eating in," to the extent that some perceive a terminal decline of the public sphere through private takeover by corporations and other agencies in the era of neoliberal capitalism. Yet not only has the ideal of familial commensality been grossly exaggerated (throughout history, including the modern period, most people have eaten their everyday meals outside the home, often alone, and generally quite quickly, since there was little to consume), the politics of the private-public distinction in the provision and consumption of food has fluctuated, depending on the prevailing articulation of technology, culture, and economy. The modern state has played a major part in regulating these forces—be it genetically modified organisms or the distilling and sale of alcohol. The point, therefore, is not to entirely disavow the distinction between the public authority of the state and the private power of markets or the household, but to recognize the regular interpenetration of these discrete spheres, and the place of politics in their changing interface: without the action of the Greensboro Four that fateful day in February 1960, Woolworth lunch counters may today still have been formally reserved for whites.

Like the moral philosophers of the Scottish Enlightenment and then Hegel, Karl Marx (1818–83) understood the emergence of a distinctive civil, commercial, or bourgeois society in the course of the long sixteenth century as the outcome of a historically unprecedented organization of social labor in the transformation of nature: the capitalist mode of production. Although characterized by the class antagonisms and social inequalities of previous modes of production, capitalism exacerbated what Marx called the "metabolic rift" between society and nature by "simultaneously undermining the original sources of all wealth—the soil and the worker."[20] Capitalism, Marx contended, widens the separation of town and country and deepens the commodification of agriculture and the consequent emergence of a class of propertyless workers, thus increasing urban squalor, but crucially also "disturb[ing] the metabolic interaction between man and the earth, i.e., it prevents the return to the soil of its constituent elements consumed by man in the form of food and clothing; hence it hinders the operation of the eternal natural condition for the lasting fertility of the soil."[21] There is, therefore, a reading of Marx that centers our attention as much upon the origins of capitalism in the process of agrarian change as on its subsequent development in urban manufacturing. Such an approach is clearly germane to our claim in this book that Marxist and, indeed, other materialist social theories are in important respects rooted in the "agrarian question": how and why have human societies moved from organizing the production and distribution of food as use-value (through communal, patrimonial, or tributary means) toward doing so mainly for exchange-value mediated through competitive markets? How, in other words, has the modern food system come to represent a market society rather than a society with markets?

No doubt, the place of food and drink in the agrarian question was for Marx and his contemporary materialists (i.e., those emphasizing the centrality of everyday structures of social reproduction in the analysis of human societies) merely a circumstantial aspect of a wider critique of political economy—the commodification of agriculture could relate as much to the cultivation of cotton or hemp as it might of sugar or corn. Yet the conceptual challenges capitalist agriculture posed for Marx's theory of ground rent or his understanding of trade, price, and inflation indicate that the industrial food system as it was emerging in the second half of the nineteenth century was a spectral presence in his work. It has since then certainly preoccupied Marxist, *marxisant,* and, more generally, historical-sociological scholarship, most obviously in the writings of the French *Annales* school and later among the advocates of dependency, world-systems, and political ecology approaches to global capitalism. Thus, as we will be discussing at greater length in chapter 10, "Political Economy," concepts like "value," "comparative advantage," "rent," or "free trade"—all derived from the eighteenth- and nineteenth-century debates involving, *inter alia,* Adam Smith, David Ricardo, Thomas Malthus, John Stuart Mill, and Karl Marx—are core to contemporary discussion of the global food system.

There is, then, no escaping the very modern problematic of the metabolic rift between society and nature or the growing tensions between culture and ecology in our understanding of the contemporary food system. The application of technology across all stages of the food chain is one obvious instance of this, and it is addressed at various junctures in the book, not just in relation to the Industrial Revolution, but also in chapters 11, "The Self," and 12, "Consumption," where the interrelationship between diet, health, cooking, and the body are shown to be strongly affected by the growing scientific engineering of nature, not least through the effects of climate change. Carbon dependence, mechanization, and genetic modification in agriculture and fisheries have incorporated these sectors—previously linked to the natural resources of land and water—squarely into the secondary and tertiary sectors of manufacturing and services, that is, into a "second nature" characterized by their full integration, from farm to fork, into the capitalist logic of value-creation. This means that—in addition to price, profit, and competition—the global food system is today subject to forms of risk, disease, and ecological degradation that are markedly different from those of premodern societies, in that they are overwhelmingly manufactured crises. Thus, modern famines are rarely the product of natural scarcity, but rather of manmade shortages. "Starvation," Amartya Sen famously argued, "is the characteristic of some people not having enough food to eat. It is not the characteristic of there not being enough food to eat."[22] Similarly, contemporary food scares—be it mad cow disease or the contribution of trans fats to cardiovascular disease—are far removed from the more basic forms of food adulteration of previous centuries, reflected instead in the positive-feedback loops inherent to what sociologist Ulrich Beck called "manufactured risk."[23]

Environmental historians too have helped us to trace the epochal changes in global food regimes triggered by complex ecological transactions. The Columbian exchange takes a leading role in our understanding of the relationship between culture and ecology when it comes to food and drink, since it captures—like few other processes—the systematic interaction between modern transformation, stratification, and globalization discussed earlier. On this account, "banana republic" is no longer just a politically pejorative label; it also conveys the reality of a global food system that, since Columbus' New World landings, has profoundly shaped the (geo)political economy of international relations. Industrial capitalism would have been unthinkable without the "ghost acreage" dedicated to sugar plantations, grain fields, and cattle ranches—as well as the enslaved labor, both native and imported—that the conquest of the Americas bequeathed the European economies. Without the introduction of potatoes, chilies, tomatoes, cassava, or avocados from the Americas, many clichéd "national" dishes in the rest of the world—fish and chips, vindaloo, *fufu*, "tricolore" salad—would have been impossible to prepare. Demographic growth across Europe, Asia, and later Africa was facilitated by the adoption of American staples as a major source of calories by peasants and workers. The "breakdown of connections" we noted earlier as marking the birth pangs of modernity has therefore found important expression in the disjuncture between nature and society in the production, preparation, distribution, and consumption of food and drink, a theme we pick up in both the chapter 2 on the Agricultural Revolution, and chapter 5, on the Industrial Revolution.

One consequence of this detachment has been the cultural, socioeconomic, and political revalorization of terroir, seasonality, and locality in response to the globalization of food. Be it the call for food sovereignty by social movements like the *Via Campesina* or more bureaucratic exercises in certifying the unique geographical provenance of certain products, food and drink have been "re-naturalized" through eminently social processes of marketing, legislating, and mobilizing for an organic, localized, artisanal, and authenticated food system. In some versions, this naturalization extends to the national, regional, or ethnic appropriation of food as belonging to—and therefore yielding a superior product within—a defined territory: proper hummus is Lebanese; real feta cheese, Greek; authentic haggis, Scottish; and so on. Paradoxically, as we argue in chapter 8, "Identity," it is globalization itself that provides a sort of global stage where different ethnicities, nationalities, and regions (or their representatives) can perform and declare their food identities. Food nationalism, we suggest, is reinforced, when not constructed, as the question "what is your national/ethnic cuisine?" invites a response that invents a coherent culinary tradition.

If spatial power and organization is the main concern of the public-private distinction discussed above, and value that of the nature-society divide, then meaning and identity are the principal categories informing our third dichotomy in the study of food, politics, and society, namely the self and other. Traditionally the preserve of social sciences like anthropology or psychology, which are especially concerned with daily structures and routines or the inner life, emotions, fantasies, and desires of human subjects, the notions of self and other appear in various guises throughout our study. They are clearly integral to chapter 11, "The Self: Food Choices and Public Health," where we consider the shifting definitions and recent medicalization of obesity in different cultural contexts across the Global North, from the level of the individual (feminism, body image) to the social (access to nutrition, food deserts). Here, the tension between Simmel's "exclusive selfishness" of eating—and the "frequency of being together" in doing so—draws attention to the interface of the personal, psychological sources of eating disorders like anorexia or bulimia and the collective, sociological dimensions of these conditions as represented in consumer culture. The complex interaction between the personal and the political is also apparent in the discussion of alcohol regulation by the state in chapter 7 as an instance of what Michel Foucault labeled "the government of the self": the production of embodied (i.e., not just psychological, but also biological) subjectivities in the political control and administration of populations as collection of bodies. Similar concerns emerge in the final chapter, "Consumption," which in many respects acts as a companion to the preceding one, "The Self." There we invoke Guy Debord's idea of society as a representational spectacle to underline the fetishized power of symbols, images, and signifiers in the mediation between food and personal identity. We are, it would seem, no longer just what we eat, but also what and how we consume. From a different ideological orientation, but sharing Debord's fascination with the semiotics of market societies, Roland Barthes wrote in an essay on the psychosociology of contemporary food consumption that "When he buys an item of food, consumes it, or serves it, modern man does not manipulate a simple object in a purely transitive fashion; this item of food sums up and transmits a situation; it constitutes an information; it signifies."[24]

Ascribing a symbolic value to food and drink is clearly not a uniquely modern phenomenon, but the cultural analysis of the meal or the sociological account of dietary prohibitions are arguably exercises in distinctively modern social theory, as is explored in chapter 4, "Culture." Indeed, the rituals associated to eating and drinking, or the cultural values underpinning different methods of preparing food, have, together with death, sex, and violence (i.e., creation, procreation, and destruction) been a mainstay of modern anthropology. Chapter 4 surveys different anthropological approaches to food and drink, showing how treating food as a universal cultural object helps us to understand the structure and variation in social and sociopolitical organization. Rituals, feasts, and festivals all express group identities and their connection to the world, both cosmologically and in relation to the rest of humankind. From James George Frazer's early reflections on totemic objects, taboos, and animistic cultures to Mary Douglas's later work on purity and danger, social anthropologists in particular have been drawn to food as an organizing signifier in the explanation of cultural practices across time and place. The expression of self that tends to emerge from such writing is one defined by the sense of belonging to one group in opposition to another, often mediated through dietary laws. Thus, as is discussed in chapters 4, 8, and 11, disgust, rejection, and prohibition in eating and drinking says less about our embodied selves and more about the collective efforts by given authorities to create a social distance from perceived others—be it on religious, ethnic, or ideological grounds. It is perhaps precisely because eating and drinking represent such "absolutely general human" physiological facts that food and drink become such powerful conduits of social identification and distantiation.

BETWEEN FOOD STUDIES AND CULINARY DETERMINISM

There may have been a time in the not-so-distant past when the pairing of food and drink with social theory would have raised eyebrows among many students of the humanities and social sciences, or at least would have been circumscribed to the specialist realms of social anthropology or psychology just mentioned. Social theory, so the perception ran, might explicitly deal with food and drink at the intimate, personal level of consumption, choice, and identity or at the very abstract level of ritual and myth. But only incidentally, or in very specialist fashion by demographers or environmental historians, on a global macro-sociological scale. This is no longer the case today—nor, we have argued, was it in the past. The relevance of social and political theory in all its variants to the study of food is nowadays uncontroversial, as evidenced by the impressive range of books, courses, journals, symposia, and conferences that discuss food politics, in both academic and wider public settings. Indeed, food studies has emerged as a self-conscious academic discipline, reflecting a growing awareness that our social lives are deeply interdependent with the planet's biosphere, and that the human need to eat and drink shapes our social structures. Our book contributes to this ongoing effort in the sociological study of food, offering a survey from the perspective of modern social theory. In this regard, it serves as a primer for those wishing to deepen their understanding of food and drink as social and political phenomena.

Yet we also aim in this volume to go beyond the mere recognition that social theory in its various forms can explain the production, preparation, and consumption of food. This introduction has staked a more ambitious claim, to be cashed out in the rest of the book, about the centrality of food and drink to the rise and development of modern social theory in the West. We argue that eating and drinking are unique in their literal, material connection between the inner world of the self and our outer social lives. These connections, moreover, have generated distinctive expressions of subjectivity, modes of production, and patterns of consumption that modern social theory has constantly returned to in its conceptualization of the public and the private; the relationship between nature and society; and the interaction between self and other. There is, to be sure, in this proposition the risk of falling into some kind of culinary determinism, whereby complex human activities are reduced to the primitive physiological fact that we all need to eat. We hope this introduction has begun to sufficiently qualify the influence of food and drink upon the development of modern social theory and has emphasized enough the powerful mediating role of states, markets, households, and civil society to allay that charge. It is in the chapters that follow, however, where the book's premise will be tested. As the saying goes, the proof of the pudding is in the eating.

CHAPTER 2

The Natural and the Social

The Agricultural Revolution

As set out in the introduction, the focus of this book is modern social theory and the way in which it illuminates the emergence of the modern food system, but also, in turn, how that system comes to shape the assumptions and concepts of social theory. An important preliminary in our enquiry is to ask about what came before the modern food system—not simply because we may be able to identify some of the key elements or developmental patterns that contributed to its rise, but also because reflections in social theory on the principal transition in human history prior to the emergence of the modern world have continued to inform its central concepts and theories. That transition is the Agricultural or Neolithic Revolution, and in the approach of social theory to it, we can see, in particular, the importance of the theme of social transformation and the organizing role played by the opposition between the natural and the social in the unfolding of human evolution. We can also discern in the agrarian history of early civilizations some of the initial features of social stratification, systems of rule and administration, and cosmological conceptions of self and other that reappear, albeit in significantly different form, in later eras. A consideration of the Agricultural Revolution thus opens the possibility of thinking about all manner of socioeconomic, political, and cultural changes and continuities in human food systems across the *longue durée*, something that the remainder of this book aims to elaborate upon.

FROM FORAGERS TO CULTIVATORS

For millions of years, hominids (the ancestors of modern humans) and early humans foraged and hunted for their food. The first signs of deliberate cultivation of plants and domestication of animals, as identified by scholars, appear about 12,000 years ago, about the time when global temperatures began to rise at the end of the Ice Age (Pleistocene), and the beginning of the current climatic age, the Holocene. In terms of the evolution of cultures, this period, the Neolithic, the New Stone Age, is distinguished from the Palaeolithic, the Old Stone Age, by a sophistication in stone tools, prior to the use of metals in the Bronze and Iron Ages. The Agricultural Revolution is thus also known as the Neolithic. Food production led to profound changes in human society, population size, division of labor, urban civilization, and complex institutions and state systems, as we shall detail in what follows.[1]

Homo sapiens appeared in the fossil record some 300,000 years ago, and modern humans in their present anatomical and behavioral forms *(Homo sapiens sapiens)* about 50,000 years ago, preceded by their near relatives, the Neanderthals, by another 50,000 years. They were the product of a long chain of evolution of hominids (proto-human) forms starting about 7 million years ago in Africa. Features of this evolution are toward upright posture, bipedal locomotion, larger body size, larger brain size, receding jaw and neck muscles, and reduced ribcage and gut size. These trends are associated with developments in material and technical cultures: evidence of manufactured tools of stone and bone, fashioned for cutting and hammering. Bipedal posture and locomotion freed the hands, allowing the fashioning and use of tools. Fossil records indicate a substantially upright posture about 4 million years ago; *Homo erectus* emerged around 1.7 million years ago, closer to us but with half the brain size; and later we see the development of language, culture, and social structure. Social formations of cooperation and coordination would also direct selection pressure toward larger brains. Modern humans, *Homo sapiens sapiens,* appeared on the scene with more sophisticated tool manufacture, more skill in hunting and fishing, and indications of more complex culture and art.[2] We know that the Neanderthal already had a complex culture, including cave paintings and ritual burial of the dead. This would indicate the presence of language, which is likely to have developed at an earlier stage of evolution. Language development would have been a great jump in cultural sophistication and social cooperation, facilitating more complex division of labor and technical innovation and, in turn, reinforcing selection pressure for larger brains and manual dexterity.[3]

Tool use had an input into the development of diet, adding to survival fitness. Cutting and stabbing tools were instrumental in facilitating meat eating, in hunting and butchering: there are indications in the fossil record of meat cut off the bones of larger animals, some hunted, but some opportunistically consumed dead animals and prey of larger animals. Meat eating was an evolutionary energy resource facilitating survival and reproduction. This was further enhanced by the discovery of fire and cooking food. Richard Wrangham has shown how cooked food facilitates ingestion and digestion by extraction of much higher levels of energy from food. This, in turn, changes the selection pressure on jaw, teeth and guts and saves energy in food gathering and digestion. Our ape relatives and earlier hominid ancestors ate a large bulk of plant matter, which required long foraging and frequent eating and the anatomical features of jaw and gut to accommodate these modes of diet. Cooking, especially with a meat component, led to more concentrated nutrition, saving on the time devoted to foraging and eating.[4]

Foraging, hunting and gathering, was the mode of subsistence of humans for much of their history. Food production, the deliberate planting and tending of crops and animals, as we saw, dates to 10,000 B.C.E. in particular regions where conditions were favorable, first in that part of West Asia known as the Fertile Crescent, which covers, roughly, parts of modern-day Iraq, Iran, Syria, Palestine/Israel, and Anatolia. Other regions developed agriculture independently, as in China; by diffusion from neighbors, as in Egypt; or through conquest by agriculture-based powers. Most hunter-gatherers were organized as nomadic bands, moving over particular territories and routes, following resources and seasons. Sustenance was primarily from gathering plant food—berries, roots, and fungi—as well as various forms of hunting, from small animals, such as grubs, mollusks, insects, rodents, and birds, to big game, under particular conditions. Fishing was another source of wild food; communities in the northwest coast of America, where plentiful supply of fish, and proficiency in conserving and storing the stocks, were provided with the conditions for a prosperous and sedentary society.[5]

A crucial factor facilitating the beginning of agriculture was the prevalence, in profusion, of plant and animal varieties that incentivized sedentary human habitation in the vicinity. Such was the case in parts of the Fertile Crescent, where seed-bearing grasses—the wild ancestors of wheat and barley—grew in abundance, making settlement possible and promoting technical inventions that facilitated deliberate cultivation, such as scythes for cutting the grasses, shaped stones for husking and grinding, baskets for carrying, and ultimately pottery. Such technical innovations must have preceded deliberate cultivation and paved the way for its adoption.[6] Pottery was the innovation that accompanied and facilitated food storage and processing: pots and vessels for cooking porridges and gruels, milking, and the processing of milk products. Wild plants also included leguminous crops, such as peas and lentils, which supplied the proteins to complement the carbohydrates from grains. But there is evidence also at this time of crop cultivation for purposes other than eating: one early domesticated plant was flax, which supplied fiber for weaving linen for clothing and was grown from around 7000 B.C.E. in the Middle East and Europe. A later crop with similar function was cotton.[7]

A more important source of protein was the domestication of animals—sheep, goats, pigs, and cattle—which followed sedentary cultivation, introducing milk products and a reliable supply of meat into the diet. Clearly, not all animals are amenable to domestication: most have characteristics that make them dangerous to humans, uneconomic in terms of diet input per food product, difficult to manage and herd, and hard to breed in captivity.[8] West Asia and the Fertile Crescent were equally favored in the animal species native to the region as the area was in the plant species. Those were the wild ancestors of sheep, goats, pigs, and, later, cattle. These are herbivores (with the exception of pigs being omnivore), which favor the ratio of feed to meat: meat-eating animals would multiply the required input of other animals that have to be fed. The chronology emerging from the archaeological record shows the first domesticate to be dogs, at 10,000 B.C.E., not widely eaten but useful in hunting and guarding. At 8000 B.C.E. emerge sheep, goats, and pigs, showing animal domestication to have occurred within the first few thousand years of sedentary agriculture. The dog and the pig were also domesticated in China. Cattle is domesticated at 6000 B.C.E. in West Asia, India, and possibly North Africa. The horse and the donkey were domesticated elsewhere in 4000 B.C.E., in the Ukraine and Egypt, respectively. There followed the camel in 2500 B.C.E., with the two varieties in Central Asia (Bactrian) and Arabia.[9] The wild ancestors of these species, in addition to being herbivores, were also amenable to herding, being of "good temper," that is, not given to territorial aggression or nervous panic (like, say, deer). Domesticated animals have utility beyond providing food material. Fertilizing manure is a valuable input for plant cultivation. Wool and skin provide material for clothing, shelters, and containers for carrying and storing. At later stages, cattle and horses are utilized for transport and, crucially, for pulling plows, a great boost to energy output.

The picture that emerges from the record is that of previously nomadic or seminomadic foragers finding areas of naturally occurring dense crops of seed-bearing grasses and legumes, incentivizing them to settle in villages and develop techniques for the exploitation, processing, and storage of these crops. Storage is crucial for food availability over the seasons. Settlement and population increase would then be a stimulus for direct engagement in cultivation rather than simple harvesting. A further step after cultivation is domestication, in which the plant develops into new forms under manipulation by humans. One direction is the selection of individual plants with larger grains. The other is the selection of plants with pods that do not shatter. Wild cereals and peas propagate themselves when the pods shatter and spill their seeds, which would then germinate and sprout. Mutant individuals do not shatter and thus retain their seeds and, in their natural state, would not reproduce. It is those mutant plants that were selected for domestication and cultivation by the early farmers.[10] Settlement and food production then facilitated the taming and domestication of animals, and it may have preceded the nomadic herding that occurred in many parts until recent times. The cultivation and domestication of plants and animals, in turn, altered the selection pressure for the evolution of characteristics favorable to human use.

Another consequence of long-term settlement was the cultivation of fruit and nut trees, which appeared in West Asia and the Eastern Mediterranean at about 4000 B.C.E. They included olives, dates, figs, pomegranates, and grapes, which were relatively easy to plant from cuttings or seeds.[11] Trees take three or four years after planting to give fruit, which makes their cultivation dependent on long-term settlement. A more difficult class of fruit trees were apples, pears, plums, and cherries, which required more complex knowledge and techniques of grafting and pollination; these were domesticated much later, probably in China.

The oldest agriculture in China, with undisputed evidence of villages, has been found in the Yellow River and the Yangtze basins, dated at about 6500 B.C.E.[12] Sites in the Yangtze basin were close to flood plains, ideal for rice. Drier and colder sites further north were hospitable to millet. These regions were home to wild ancestors of rice and millet, just as Southwest Asia was of wheat and barley. Cultivation, then domestication, involved the same process of selection of non-shattering seeds as was found in the Southwest Asian examples. One interesting difference between the two regions was that pottery appears to have preceded agriculture in the East. Sites in southern Japan and China contained pottery that was dated as early as 12,000 to 9000 B.C.E., long before the start of agriculture.[13] This can be considered an indication of the prevalence of a cuisine based on boiling rice and millet, indicating an abundant supply of these plants before domestication. This was partly contemporary with flour-based cuisine, indicated by the presence of large numbers of grinding stones found in late Palaeolithic and early Neolithic Chinese sites. The archaeological record indicates a continuation of hunting and gathering activity in parts of Japan, China, and Korea well after the beginnings of agriculture, which would indicate a reluctance of affluent foragers to engage in cultivation.[14] The rigors of rice cultivation, which were to become an important element in the social and political organization of the region, may have been the reason for this reluctance.

The staple crops considered so far in relation to the development of agriculture have been primarily cereals and legumes. Some regions, parts of the Americas, Africa, and New Guinea, relied more on tubers— yam, manioc, and potato—and squashes.[15] One advantage of cereals and seeds over this category of crops is amenability to storage. Dried cereals, as we saw, can be stored in quantity over long periods, and surpluses can be used in a complex economy of exchange, taxation, trade, and even as means of payment, a proto-currency. The more perishable tubers lack these qualities. This is another aspect of the early advantages enjoyed by the cereal-based economies of the Fertile Crescent, Egypt, and China.[16]

EFFECTS OF CLIMATE AND TERRAIN

We saw that the Fertile Crescent was especially favored because of its naturally occurring high concentration of wild plants and animals to feed former foragers and induce them to settle, cultivate, and domesticate those varieties. Another factor in its favor was the climate and the terrain, which also facilitated the diffusion and spread of agriculture within its climatic zone, that of the Eastern Mediterranean. It is characterized by mild, wet winters and long, hot summers. The plant dies during the dry season, while the seeds remain dormant and spring to life once the rains come.[17] The plant remains a small herb while its energy is directed to producing big seeds, the means of its perpetuation.

The diffusion of food production and its repertoire of plants was facilitated by the wide area enjoying this Mediterranean climate, from northern Mesopotamia through the Levant, Anatolia, and into the Balkans, as well as western Iran. Animals and plants domesticated in one location could thus be successfully transferred within that area in a relatively short space of time. Sheep and goats were probably first domesticated at high elevation in the Zagros mountains in Iran or in the Levant, cows probably in Anatolia, but all soon became common throughout the region. The diversity of the terrain also contributed to the spread. It included mountains and highlands as well as river plains and deserts. It was thus possible for early cultivators to take seeds from high altitudes, dependent on unreliable rains, down to river valleys with regular floods and irrigation.[18]

Jared Diamond advances an interesting account of the geography of diffusion of food production in the different continents. He contrasted Eurasia with both the Americas and Africa in terms of the main axes: east-west in the former, north-south in the latter. The first comprises regions on similar latitudes, which thus have similar climatic types and rotation of seasons, facilitating the diffusion of repertoires of plant and animal domesticates. In the case of the Americas and sub-Saharan Africa, such movement is more difficult, with climate and seasons ranging from the equatorial to the Mediterranean climate of southern Africa and California. Along this north-south axis, there are also major geographical and climatic obstacles, such as deserts and tropics, separating regions of similar climatic regimes, such as Mediterranean North Africa and the south of the continent, as well as the Andes and North America. This would explain the relatively rapid spread of food production from West Asia westward and northward to Europe and eastward to India, as well as the spread of Chinese agriculture to Japan, Southeast Asia, and Oceania. Diamond refers to geneticist Daniel Zohary and botanist Maria Hopf on the pace of the spread of food production from West Asia to different parts of Eurasia.[19] Starting about 8000 B.C.E. in West Asia, it reached Greece and parts of India by 6500 B.C.E., Egypt about 6000, central Europe by 5400, southern Spain by 5200, and Britain around 3500. By the beginning of the Common Era, Fertile Crescent domesticates were cultivated from Ireland to Japan. In contrast, domesticates from one part of the Americas failed, or were slow, to spread to other regions. Diamond gives the example of the failure of domesticates from the South American Andes, such as llamas, guinea pigs, and potatoes, to spread to the highlands of Mexico, which had ideal conditions for their cultivation. The spread between these two highland regions was prevented by the intervening hot lowlands of Central America. The conditions that facilitated the spread of agriculture and the repertoire of domesticates also favored the transfer of technologies and materials. The wheel, for example, invented in West Asia around 3000 B.C.E., spread east and west across Eurasia within a few centuries, whereas the wheel invented in Mexico did not spread to the Andes.[20]

HUNTER GATHERERS

To trace the social transformations following agriculture, let us first examine the mode of life of hunter-gatherers who preceded that revolution. The Agricultural Revolution spread to many parts of the world and became the basis for civilizations and empires but did not entirely eliminate foraging economies. These survived in many parts and were still prevalent in early modern times, on the fringes of agriculture-based complex societies. The Industrial Revolution, the world market, and, crucially, their spread through colonial domination, led to the incorporation and marginalization of many remaining forager societies. A few, however, survived into the twentieth century, notably parts of the Australian native population, the !Kung Bushmen in southern Africa, and some Native Americans. These were, then, subject to anthropological studies and travel and missionary reports from the eighteenth century on—some of which we shall be considering in chapter 4. These studies, as well as archaeological records, have been the sources for the characterization of the hunter-gatherer societies. There are, of course, a diversity of hunter-gatherers, depending on habitat and climate, but some common characteristics are discernible.

Many hunter-gatherer societies are organized as small bands, typically nomadic, moving from one camp to another following the seasons and the availability of food items. Their movement follows set cycles. The nomadic life and the limited resources do not allow for growth of population, and if a band gets beyond a certain limit, it is likely to split into two or shed members through marriage to other units. The nomadic life sets limits to the number of children that any woman can carry and nourish, with births being well spaced. Various measures are typically taken to that end, such as sexual abstinence, abortions, and even infanticide. The division of labor is simple, what was called "natural," by gender and age. Generally, women are the gatherers and men the hunters. Here is an excerpt from a description of two Australian groups studied in 1960:

[On Groote Eyland] there were four adult males, four adult females, and five juveniles and infants in the camp. Fish Creek was investigated at the end of the dry season when the supply of vegetable foods was low; kangaroo hunting was rewarding, although the animals became increasingly wary under steady stalking. At Hemple Bay, vegetable foods were plentiful; the fishing was variable but on the whole good by comparison with other coastal camps visited by the expedition. . . . In inland hunting, good things often come in large packages; hence one day hunting may yield two days' sustenance. A fishing-gathering regime perhaps produces smaller if steadier returns, enjoining somewhat longer and more regular efforts.[21]

Marshall Sahlins, in a notable essay, argued that hunter-gatherers were the "original affluent society." The polemic was directed against common characterization of foragers' conditions of existence as harsh and precarious, in a constant quest for food and exposed to the elements in difficult environments. In the famous words of the political philosopher Hobbes, in "man's natural condition," life is "solitary, poor, nasty, brutish, and short."[22] Sahlins argued, to the contrary, that foragers had few needs, and those were easily accommodated within their means. Food, he argued could be procured in plenty in a few hours or days of work, as we see in the above quote on Australian hunters: "one day's hunting may yield two days' sustenance." And given that food could not be stored, all that was procured was eaten within a short interval and, when profuse, shared. Sahlins quoted earlier reports, from mid-nineteenth century commentators, which relate even more spectacular abundance: "I have seen six hundred natives encamped together, all of whom were living at the time upon fish procured from the lake, with the addition, perhaps, of the leaves from the mesembryanthemum. . . . At Moorunde when the Murray annually inundates the flats, fresh-water crayfish make their way to the surface of the ground . . . in such vast numbers that I have seen four hundred natives live upon them for weeks together."[23] He goes on to add that those same "natives" procured a kind of cress from cavities in the mountains at the appropriate season. Such abundance may have been exceptional for place and season, but for Sahlins, at the high end of common availability of food to hunter-gatherers, making them, within the limits of their existence, the original affluent society. Women engaged in gathering plants accumulated plenty with, typically, a few hours of work per day. They spent much time in socializing and sleep. Beyond food, the foragers had few needs: rudimentary clothing, tools for hunting and digging, and a leather bag for carrying. In a nomadic life, possessions are a burden and are kept to a minimum: only one item of most possessions, such as bags and tools, and a few arrow tips. Rudimentary shelters were constructed at each camp. Sahlins found these conditions described in studies of various hunter-gatherers in Africa, the Americas, and Oceania.

Sahlins's characterization has, of course, been challenged by many examples that may not fit. Yet his interesting accounts illustrate a common theme: that foraging sustained an easier and more leisurely, if precarious, mode of existence than the constant discipline and work required by cultivation. For many communities in different parts of the world, there was no definitive, evolutionary leap from foraging to cultivation in the Agricultural Revolution, as may be suggested by the term *revolution,* but partial steps, leading to various combinations of the two, and even occasional reversion.[24] The comparative ease of foraging as compared to the rigor of agriculture may account for the examples of foragers existing in proximity to agricultural communities but not taking up cultivation, of others who revert to foraging, and of those who combine the two. Graeme Barker argues that husbandry—manipulating the environment, plants, and animals—is as old as the beginning of humankind. Some hunter-gatherers intensively manage their land: New Guinea people who never domesticated sago palms, nevertheless increase production of these and other wild plants by clearing other competing plants and keeping channels in the swamps clear.[25] Other examples are the burning of landscapes to enhance production or clearing and tending land where food is foraged, as Australian natives did for wild yams. Slash-and-burn agriculture, a form of seminomadic cultivation, continued in parts of Africa till recent times. Areas of the forest are cleared by burning, then cultivated for a few seasons until their fertility declines, at which point the seminomads move on to other locations. Barker quotes examples of communities that combine intensive agriculture with foraging. The Kelabit of Sarawak engage in wet-rice agriculture, and rice plays a central role in their diet and their culture, bestowing much status on its successful cultivators. Yet much of their diet also comes from the foraging of animals and plants from surrounding forests.[26] Agriculture, then, argues Barker, was not a revolution or a decisive leap, but it proceeded over much of human history in various patterns and combinations of foraging and production.

Slash-and-burn agriculture is just one example of seminomadic cultivation. Apache Indians in North America farmed in the summer at high elevations, then moved to lower terrain further south to search for wild foods during the winter. Diamond also cites the example of modern nomads in New Guinea who make clearings in the jungle, plant bananas and papayas, wander off for months of foraging, and then return to settle for a while and eat their crops.[27] We also find examples of sedentary hunter-gatherers, especially in coastal fishing communities. North America's northern Pacific coast was host to such a community, who became sedentary but not food producers. Coastal Peru and Japan became sedentary long before they turned to food production. The regular and abundant supply of fish and techniques of preservation and storage provided the basis for permanent settlement.

CONSEQUENCES OF THE AGRICULTURAL REVOLUTION

The shift from nomadic foraging to settled cultivation had profound consequences for the societies thus affected. The first was demographic: settlement allows more frequent births. We saw that nomadic foragers have to restrict the frequency of childbirth, because of the difficulty of carrying children on the move. This constraint does not operate on sedentary farmers. The birth intervals for many farmers is about two years, half of what it is for foragers.[28] This higher birthrate, as well as a larger and more secure supply of food, leads to much higher population densities in agricultural societies.

Food storage is another feature of settled agriculture. We saw that, with few exceptions, foragers cannot store or carry food, even when a surplus can be procured. Cultivators, settled in permanent sites, could produce and store surplus food, to be consumed over the year, but also, crucially, to feed others who are not engaged in food production. Wheat, barley, and legumes in the Fertile Crescent, and rice in East Asia, are perfectly suited for long storage. This has revolutionary significance, for it allows a more complex social division of labor and, ultimately, the development of administrative, political, and military power structures. One of the founders of political sociology, Max Weber (1864–1920), began thinking about his canonical distinction between three "ideal types" of legitimate rule with reference to studies on agrarian relations in antiquity, associating the birth of cities, aristocracies, taxation, armies, bureaucracies, and later commerce to the "joint efforts to produce food" and the "special allotments of land."[29] At the simplest level, surplus production allows the development of crafts and of administrative and coordinating functions, including those of irrigation, storage, and distribution. Food storage also raises issues of security: the protection of stores from rivals and marauders. Thus it gives rise to military formations, at first in the form of a "peasant army," able-bodied members in arms, but ultimately professional armies. These are typically linked to political authority and bureaucratic administration. In many such societies, such as those of Mesopotamia—Babylon and Assyria—we also have the development of elaborate religious institutions, with a priestly caste. These have the function of legitimizing authority, but also of codifying and deploying social knowledge and technique with respect to the seasons, astronomical calculations, and literacy. All these classes of non-producers are sustained by systematic taxation of the producers. The production of agricultural surpluses as well as crafts fosters trade, both internally within the administered territory and externally with neighboring or even distant communities. We see trade occurring between Mesopotamia, Anatolia, Iran, and even Egypt. Trade, in turn, also raises issues of protection, thus expanding the military functions. All these developments involve the formation of hierarchies of power and wealth, and thus classes, inequalities, servitude, and exploitation.

There is an interesting fictional illustration of some of these processes of stratification and appropriation in two classic films, the Japanese *The Seven Samurai* (1954) and the Hollywood remake *The Magnificent Seven* (1960). A village is subject to periodic raid by gangs of bandits, who loot the stored grain. The village leaders, unable to mount adequate resistance, approach a group of samurai to defend them, offering them a share of the stored food. The samurai undertake the task with great success, ultimately killing the bandits. We can imagine a sequel, which is all too common, of the samurai then settling down to this protection on a regular basis, sustained and financed by taxation of the peasants, thus setting up what would become a feudal society. This fictional example is one of rudimentary and precarious formation of hierarchical political/military entities. The complex formations of the early cities and states of Mesopotamia proceeded over several millennia from the first stages of settlement and cultivation; the first small city-states in Mesopotamia are dated to about 3000 B.C.E., some four millennia after the first cultivation.[30]

In contrast to foraging, agriculture imposes, in most instances, a work discipline, requiring constant effort in relation to the seasons and cycles and stages of growth of the crops and the tending of trees. The leisurely life of the forager, as portrayed by Sahlins, is not for the peasant. In addition, in the more successful and favored locations, food production sets in motion social, economic, and cultural processes that led to complex societies and polities—indeed, to "civilization," as we shall see presently. These societies, however, are typically stratified by class, with the actual producers, the peasants, being exploited, having to render much of their production in taxes and rents, sometimes in conditions of servitude as serfs or slaves. In addition, the high population densities are conducive to the spread of infectious diseases. The proximity to animals is an additional factor in infections and epidemics.[31] It is in contrast to such regimes that Sahlins's designation of foragers as the original affluent society makes sense. Agriculture-based societies and economies covered much of the world by early modern times and were of vastly different forms in scale and organization. Many such societies, termed *stateless,* were founded on household production and kinship-based social organization, some so-called tribal. Much of this is subsistence or near-subsistence agriculture, that is to say, production for direct consumption. Authority and coordination functions, as well as exchange of food and commodities, are "embedded" in those social relations of kinship and marriage.[32] Chiefdoms and various forms of state societies develop with enhanced production, exchange, security and coordination functions, tribute, and taxation. There are diverse forms of state societies, in terms of institutions of authority, forms of distribution and exploitation of agricultural surpluses, and forms of labor organization, from household production to slave or other forms of forced labor on large estates (latifundias). Let us consider here the example of the earliest agrarian civilization that developed in the Fertile Crescent and created the first cities as centers of government, economy, and the innovations that defined subsequent civilizations, notably writing.

There are various hypotheses about the early development of complex forms of government in Mesopotamia. Gordon Childe,[33] one of the earliest archaeologists to advance a narrative for those developments, stressed irrigation systems in river valleys as an important factor creating functions of management and coordination, with attendant differentials of power and knowledge. The alluvial basis of these early civilizations (i.e., their dependence on the natural fertilization of land through silt and mud deposits from flooding rivers) inspired environmental historian Karl Wittfogel's thesis of ancient "hydraulic empires" built on the control and administration of irrigation.[34] Though since discredited for its environmental determinism and geographical rigidity, Wittfogel's emphasis upon alluvial agriculture has nonetheless combined with Max Weber's sociology of domination to deliver a renewed historical and comparative sociology that has the distribution of agricultural surplus at the core of their theories. Michael Mann's monumental investigation in his *Sources of*

Social Power is perhaps the most influential expression of this approach.[35] In the first volume of that work, Mann focuses on the Mesopotamian experience and invokes the idea of early civilizations as "cages" that "insulate" society from nature through institutions like ceremonial centers, writing, and cities. "The decisive feature of these ecologies"—the alluvial states along the Indus, Nile, Tigris-Euphrates, or Yellow River basins—"was *the closing of the escape route*. Their local inhabitants, unlike those in the rest of the globe, were constrained to accept civilization, social stratification and the state. They were trapped into particular social and territorial relationships, forcing them to intensify these relationships rather than evade them. This led to opportunities to develop both *collective* and *distributive* power."[36]

The organization of labor for digging and maintaining canals, the regulation and rationing of water flows, and knowledge of seasonal and climatic conditions, all bestow considerable power on upper classes: managers, chiefs, soldiers, and experts, the latter institutionalized as priests. These administrative functions then create the need for record keeping, accounting, and the storage of knowledge of the natural cycles, as well as the religious ideologies in which they are embedded. All these functions provide conditions and needs for writing—a momentous innovation in the history of humanity—for recording and transmission, first of administrative and economic records and calculations, then ever more elaborate bureaucracies of power and administration, taxation, and trade, and eventually much more generalized functions of religion, literature, and even cooking manuals. We shall turn to the evolution of writing presently.

First, we should note another momentous development accommodating the classes and functions thus created: cities. The production of surplus food and artifacts, of pottery, metalwork and clothing, lead to networks of exchange, becoming markets with standardized units of value: money. Cities develop from the aggregation of administrative, religious and commercial functions and their personnel: palaces for the rulers, temples, markets and housing. Cities are, then, the centers of "civilization," and the innovation of writing the marker of this achievement. Evolutionary theorists from the nineteenth century, such as Herbert Spencer and L.H. Morgan, postulated stages in the development of humanity, from "savagery," corresponding to foraging and the old stone age, to "barbarism," marked by the Neolithic/Agricultural Revolution, then "civilization" marked by cities and writing. With writing, pre-history becomes history. City-states, in turn, involved in expanding networks of trade, war and conquest, develop into kingdoms and empires. Mesopotamia, the earliest theatre for the formation and succession of states, dynasties and empires, we have Ur, of biblical fame, in the south of present day Iraq, with the dynasty of Lagas ruling a city state in 3000 B.C.E. Later waves of migrations, conquests and settlement bring in ever larger units, of Akkadians, Sumerians and Assyrians, centered in cities such as Babylon and Nineveh, continuing through from 2300 to 500 B.C.E. and the Persian conquests, then Alexander and the Hellenic period from 330 B.C.E.[37] One "golden period" was the unified Babylonian empire under Hammurabi (1792–1750 B.C.E.), the first Middle Eastern "law-giver" preceding and anticipating the Bible and the Qu'ran.

Sumerian and Akkadian writing in Mesopotamia is dated from about 3000 B.C.E., the earliest or joint earliest with the Egyptian (though it is not clear whether the latter developed independently). Mexican Maya writing is dated to 600 B.C.E., the only other indisputably independent development. Chinese, which may have been independently developed, was in 1300 B.C.E. The Mycenaean Linear B, the earliest form of Greek, also perhaps developed independently, in 1450 B.C.E. All other writing systems were derivative, either by adoption of an existing system or the idea of writing inspired by example.[38]

Sumerian writing, in the style called *cuneiform,* consisted of characters imprinted on clay tablets that were then dried. It evolved from the rudimentary clay tokens used by villagers and administrators to indicate quantities of produce or animals to more complex writing—using reeds to inscribe symbols into the clay—to indicate quantities and persons and to record transactions and obligations, such as taxes and debts. More elaborate notations followed, of logograms, initially pictorial then more abstract, some combinations used to indicate phonemes, sound effects/units. With this evolution we have the enhanced forms of expression, from simple record keeping to elaborate texts of religious, scientific and poetic expressions, and including cookbooks and recipes, an example of which will be considered below. Egyptian writing, on papyrus sheets, Chinese, on paper, and Mexican, on stone, also consisted of logograms, combining pictorial, symbolic, and phonetic elements. Alphabetic writing, in which words are formed from phonetic letters, was first developed by the Phoenicians in the Mediterranean Levant, circa 1050 B.C.E., in what is now Syria/Lebanon/Israel/Palestine, and as sea-faring peoples, they carried it to various corners of the Mediterranean, where they established trading colonies. Phoenician writing was the basis for other Semitic languages, such as Hebrew, Aramaic, and Arabic. It was also adopted by the Greeks, who added the further elaboration of vowels, while the Phoenician and other early scripts had consisted of only consonants, with diacritical signs indicating vowel sounds, as is still partially the case in Arabic and Hebrew.

A MESOPOTAMIAN COOKBOOK

Jean Bottero, a French archaeologist, wrote *The Oldest Cuisine in the World: Cooking in Mesopotamia,* based on his research of collections of clay tablets relating to food, drink, and cooking from different periods and locations, starting from the second millennium B.C.E., written in one or both of the languages of the time, Sumerian and Akkadian. The latter is a Semitic language, related to later Hebrew, Aramaic, and Arabic, and it is interesting to note some affinities between some Akkadian food words and say, later Arabic usage.

Bottero distinguished indirect sources, those related to food materials, from the direct texts on eating, cooking, and recipes.[39] The first category includes itineraries of materials in palace archives and correspondence between individuals. An interesting element in the indirect sources are tablets that specify payments to individuals, like wages, in quantities of grain, indicating that grain functioned like money. The grain thus dispersed is, presumably, not just for personal consumption but also for use in other payments and purchases. This use of units of food as payment recurs in other agrarian societies: in Japan, until early modern times, samurai received payment in units of rice; the word *salary* is derived from the Latin word for "salt," indicating its origin in payments in units of salt. Other indirect sources include records of palace stores and accounts that list a range of different kinds of cereals and flours (staple ingredients); fish (maybe dried and salted); different cuts of meat, fat, and organs from sheep and cattle; different birds; oil; honey; and, crucially, beer, to which we turn presently. The correspondence is between relatives and discusses business transactions, specifying a range of goods, including varieties of fish and different types of onions and garlic. Another class of texts relates to what Bottero calls "divinatory treatises": the prediction of future events through oracles and dreams, often involving animals and food. For example, if a man dreams that he is eating the meat of wild animals, disaster may befall members of his family. Some tablets have been described as "encyclopedias" of the material world, listing, for instance, trees and plants, and also types of food. Two hundred varieties of breads are listed, differentiated by the flours used, the kneading, additives, and flavors. The encyclopedias also list fifty dairy products and cheeses.[40]

Bottero's direct sources are those tablets that describe the composition and preparation of dishes.[41] Preparations reveal the *batterie de cuisine*, the equipment and pots. The main cooking fuel was wood and kindling, possibly hay, burned in stone hearths built in the ground, over which clay pots and possibly metal kettles were placed. More elaborate was the *tanuru*, a clay oven built into the ground with an opening at the bottom for burning fuel, used primarily for baking, by sticking rounds of dough on its heated walls. The *tanuru* was also used for the slow cooking of meat in sauces placed at its mouth. This is the ancestor of the modern-day *tannur (tandir/tandur),* used in the Middle East and parts of India, in similar operations. It shows a high degree of sophistication.

Now to the cooking. The simplest would be meat roasted directly on the fire. This, apparently, was only or primarily used for meats offered ritually to the gods. It was not the Biblical "burnt offerings" which was not a Mesopotamian ritual—in Mesopotamia, the gods were believed to like real food. The main form of cooking meat and vegetables was in water mixed with fat along with garnishes and spices. Typically, the cook is instructed to "prepare water, add fat, mashed leeks and garlic."[42] As well as various meats, the broth could be thickened with crumbs or *samidu,* a fine flour, then a repertoire of spices, including cumin and coriander. Blood and milk were occasional additions. A range of vegetables are featured in the recipes. Stewing meats and baking a variety of breads and cakes appear to be the dominant cooking methods, but there are also instructions for pickling, salting, and fermenting fish, meat, and vegetables. This was a relatively sophisticated cuisine for an early development of agrarian civilization.

Drinks, other than water, mentioned in the tablet texts are milk—a minor item, given its perishability—and most important, beer.[43] Beer seems to have been a dominant item in Mesopotamian cultures from the earliest times, and certainly on the record since the beginning of writing at the end of the fourth millennium B.C.E. These ancient preparations shared the basic process of beer making that we know, though the final product differed markedly from our present-day beer. This basic process used various grains, of which barley was the most common, germinated and malted in damp conditions, then heated in water to which aromatic elements were added (though not the hops used today, which was then unknown), then left to ferment.[44] It was mostly produced domestically, by women, and was typically consumed directly from the tub in which it fermented, through straw-like tubes. Beer was regularly consumed at meals but was also drunk recreationally. It was celebrated in poems and songs for its taste and intoxicating effects. In addition to domestic consumption, it was offered in "taverns," first by people offering their surplus domestic production, then in specialized premises.[45] Authorities were concerned about possible disorder on these premises, whether of moral infringements or political subversion, a worry, as we will see in chapter 7, that has persisted in many later civilizations, including some contemporary examples. Beer was also an essential and regular item in temple offerings to the gods. A record of such offerings enumerates many types of beer, with different colors and flavors, indicating sophisticated discernment.[46] Wine was known but not commonly consumed. It was not locally produced on the plains where the principal cities were located, which were not well endowed with trees and vines, but it was imported from the uplands of Syria, transported by boats on the Euphrates.[47]

We should note that fermented alcoholic drinks, whether made from cereals or fruit, were common features of all the early agrarian civilizations, such as Egypt and China. Alcoholic drinks feature prominently in the symbolic cultural universes of all those societies, with legends, rituals, and poetry. Decorated drinking vessels, to be seen in archaeological collections in museums, bear witness to these symbolic functions.

CONCLUSION

This chapter has narrated the early development of food systems and how it interacted, first with biological evolution and then with the formation of society, polity, and culture. We saw how diet and cooking were instrumental in directing selection pressures in the evolution of *Homo sapiens sapiens*. Then, the domestication of plants and animals by humans in the process of food production directs selection pressure in the evolution of those species. Moving from biology and nature to society and culture, we saw how means and patterns of subsistence entered into shaping social relations and organization, first in hunter-gatherer societies and then in the profound and varied transformations effected by food production in the Agricultural Revolution. We then concentrated on the particular line of such development that led to demographic growth and structural complexity, with the production and distribution of food surpluses and the complex division of labor, power, and knowledge that it made possible. We traced these processes in the example of Mesopotamia and the development there of the earliest cities, writing, and civilization. Part of that process was the development of cuisine and of the culture and lore of alcoholic drinks, read by archaeologists in cuneiform texts on clay tablets.

The scene is thus set for the complex and manifold developments to follow in the history of society, polity, and culture, the subjects of the chapters to follow. In particular, early processes of social stratification, urbanization, political administration, commerce, and cosmological interpretation have been shown to have emerged from specific conceptions and interactions between nature and society. In the next chapter, we consider how this interrelationship unfolded through the European conquest of the Americas and the beginnings of a properly global interdependence between diverse societies across the planet. This, in turn, opens the way for a closer examination in later chapters—especially those dealing with industrialization and political economy—of how the modern shift to a carbon economy and a capitalist mode of production, beginning in the eighteenth century, transformed not only many of the socioeconomic, political, and cultural infrastructures inherited from the Agricultural Revolution but also the very relationship between the natural and the social that had prevailed until then.

CHAPTER 3

Exchange

The Columbian Exchange and
Mercantile Empires

To begin, some trivia: in 1998, a song entitled "Vindaloo" became the unofficial anthem for English fans at that year's FIFA World Cup in France.[1] The reasons behind the choice of a Goan dish as a signifier of English soccer prowess remain obscure—although some might consider it a reflection of the country's assuredly postcolonial status, much in the same way that the generic tikka masala became the United Kingdom's favorite comfort food at the start of the new millennium. Some fifteen years after that World Cup, a restaurant called "Ceviche" opened in London's Soho, launching a new trend in the British capital for Peruvian food and the famous cocktail pisco sour, which has grown unabated since then.

These two seemingly random anecdotes are germane to the themes of this chapter because they reflect the impact of imperial conquest and exchange on modern foodways. Both the dishes mentioned (and, indeed, also the drink pisco sour) contain ingredients that were alien to the peoples and places we now call Peru and Goa. Citrus fruits and grapes were brought to the New World by Spanish conquistadors, and chilies would have been unknown to South Asians until the Portuguese introduced them to their eastern colonies from the Americas. Ceviche—slices of raw fish or shellfish marinated in lime juice, chilies, olive oil, tomato, and onion—is associated to the Spanish word *escabeche*, describing a food preparation where (cooked) fish is also pickled in vinegar, onion, sugar, and spices. *Escabeche* has, in turn, been traced to the Arabized

Persian word *al-sikbaj,* referring to fish or meat cooked in vinegar.[2] Based on this etymology, it is likely that the Peruvian ceviche is a Latin American version of the Middle Eastern *al-Sikbaj,* which made its way westward through the Muslim invasions of Iberia from the eighth century C.E. and subsequently via the sixteenth century Spanish conquest of the American mainland. Similarly, the word *vindaloo* is a vulgarization of the Portuguese *vinho e alho* (wine—or vinegar—and garlic), two of the key ingredients for the pork- or chicken-based specialty of Christians from India's west coast.[3] It is an outcome of the eastbound trade in New World produce created by the Portuguese and other early-modern colonial empires.

These various linguistic and culinary trajectories tell a story of movement facilitated by imperial expansion, which the environmental historian Alfred W. Crosby Jr. in 1972 famously labeled the "Columbian exchange"—the immediate sociobiological, botanical, and ecological consequences of contact between Europe and the Americas after Christopher Columbus's landfall on Caribbean soil in October 1492.[4] As chapter 2 has already shown, imperial trade in food staples and the accompanying intercontinental transfer of agricultural and horticultural techniques long predates the European conquest of the Americas. There is, similarly, no denying Gary Paul Nabhan's claim that historical trade routes like the Silk Road and multiple other cross-regional commercial, caravan, and pilgrimage networks had previously pioneered new foodways and disseminated their accompanying socioeconomic structures and processes across land and sea.[5] But the truly global scale of the Columbian exchange, allied to its momentous contribution toward the industrial organization of the modern food system along capitalist lines, warrant calling it an instance of epochal or world-historical change.[6]

The Columbian exchange did not, however, simply involve the cross-Atlantic trade in goods (corn, beans, and squashes from one end; livestock, sugar, and wheat from the other). It was, above all, a *colonial conquest:* that is, a systematic attempt at subjecting whole populations and their resources to the rule and benefit of another (generally distant) population, with all the consequences such forms of domination have for socioeconomic relations, political structures, and ecological systems—both at home and abroad.[7] Moreover, an unintended result of the exchange was the integration of the peoples and resources of the Americas into a worldwide imperial system of circulation, appropriation, and exploitation—characterized by, among other things, plantation slavery and the accompanying "triangular trade" in humans, stimulants, and weapons—which, according to Karl Marx, served as the foundational moment in the "previous" or "primitive" accumulation of capital.[8] From the sixteenth to the nineteenth centuries, over ten million Africans were forcibly transported from their homelands to the New World, mainly to produce sugar, tobacco, coffee, and cotton for European markets. Millions of Amerindians perished—some aboriginal populations wiped out entirely—in the epidemics that ensued after initial contact with Europeans. We should therefore avoid a narrow, purely descriptive notion of "exchange" that obviates the enormity of the human sacrifice that accompanied the European age of discoveries.

After outlining the main features of the Columbian exchange, the rest of this chapter considers processes which the early twentieth-century economist Joseph Schumpeter (1883–1950) called "creative destruction":[9] the radical transformation of existing ways of life (and death) by socioeconomic and political forces that uproot previous social structures and natural environments while simultaneously transplanting and reconfiguring new ones, often adapting and adopting aspects of preceding systems. As we shall see, the Columbian exchange occasioned the catastrophic collapse of Amerindian populations and their civilizations, while building distinctive new Creole societies and institutions that in many important ways drew on pre-Columbian social structures and organizations. It also integrated vast, previously unconnected areas of the planet into a world market that generated considerable wealth and increased living standards at the same time that it reinforced and, indeed, invented some of the most brutal forms of social domination and segregation. Put bluntly, the Columbian exchange was a prime example of creative destruction.

The production, preparation, and consumption of food and drink was central to this process, and so, in what follows, much attention will be placed on their role in reproducing colonial empires, particularly in the new societies of post-Columbian America. Specific attention will be paid to mercantile empires of the early modern period, which derived their national wealth principally through the exploitation of conquered lands and foreign bonded labor, combined with the naval control over the world's most lucrative maritime trade routes. This historical focus will, finally, pose important questions, addressed toward the end of the chapter, about the imperial legacies in the present, supposedly postcolonial period, when notions of authenticity, hybridity, and cosmopolitanism in relation to food and drink are inescapably connected to the era when empires acted as the main facilitators of international traffic in goods and peoples. In order to make full sense of these complex phenomena, we need to draw from different branches of the social sciences and humanities identified in this book—environmental history, social anthropology, demography, and postcolonial and area studies, as well as history and political economy—that have provided some key concepts in the explanation of how and why chilies are now consumed in South Asia and lime juice in Latin America and, more importantly, what the consequences of such exchanges are for the modern interaction between food, politics, and society.

THE ENCOUNTER

Spaniards arrived on American shores in 1492 searching for a maritime shortcut to the Spice Islands of the East Indies. Columbus went to his grave with the firm belief he had discovered a new route to East Asia, not the Caribbean. But early contact with gold in the Antilles led Spaniards to the infamous hunt for El Dorado (the mythical tribal chief who sprinkled himself with gold dust as part of an initiation rite). The subsequent encounter with mainland American civilizations rich in silver and gold simply exacerbated what was to become one of the main drivers for the European conquest of the new continent. In this quest for bullion, Spaniards brought with them crops, animals, and technologies that would not only feed European conquerors and settlers but also extend the frontiers of Christendom (as represented by the Spanish Catholic monarchs and subsequently by the Habsburg and Bourbon dynasties) into American territories unabashedly given names like the Vice-Royalty of New Spain, New Granada, or River Plate. Thus sheep, goats, cows, pigs, horses, donkeys, and mules (all ungulates—herbivores with hard hooves) as well as olives, mangoes, sugarcane, wheat, coffee shrubs, and banana trees, among scores of other plants and animals, were introduced on the American continent through European conquest. In exchange, Europeans brought back to the Old World (both Eurasia and Africa) American produce that had never been cultivated outside the western hemisphere: potatoes, sweet potatoes, squashes, tomatoes, peanuts, quinoa, pineapples, chili peppers, corn, tobacco, cacao, vanilla, and avocados, among countless others.[10]

The word *exchange* suggests that this was a reciprocal and relatively free interaction, but the unevenness in the form, content, reach, and pace of the Columbian exchange already points to the inequality inherent in the process. For a start, as Rachel Laudan has indicated, the transfer in food processing techniques and forces of production—including beasts of burden—went principally in one direction.[11] Old World agriculture and pastoralism conquered New World horticulturalism, but the opposite was not the case; vineyards and wine accompanied European settlers, but agave and its alcoholic derivative, *pulque,* did not extend across Eurasia or Africa. While many Native American peoples soon mastered horsemanship and adopted pork, chicken, beef, and mutton as sources of protein, Europeans did not import any American ungulates (alpacas or llamas) to the Old World's food or transport system, nor did they— with the important exception of the turkey—introduce other New World meats like guinea pigs or iguanas into their diet. Of all the indigenous American agricultural techniques, only the bird-dropping fertilizer *guano* was enthusiastically adopted in Europe—and that only for a short period during the late nineteenth century. Moreover, the American foods, plants, and stimulants that did go on to have a phenomenal impact on the rest of the world—corn (maize), potatoes, sweet potatoes, cassava, chocolate, and tobacco—spread comparatively slowly, and certainly irregularly, across Africa, Europe, Asia, and beyond, generally taking centuries to become local staples.

The case of the potato *(Solanum tuberosum)* is an instructive and controversial one, as scholars dispute the speed at which it was disseminated from its origins in the Andean highlands. Most recent accounts confirm Redcliffe N. Salaman's claim in his classic study, *The History and Social Influence of the Potato,* that the first recorded instance of the tuber's consumption in Europe appears in the accounts of Seville's Hospital de la Sangre, where in December 1573 nineteen pounds of potatoes were supplied to feed hospital patients.[12] Fernand Braudel's own monumental study on everyday life in early-modern Europe also refers to late sixteenth-century texts that mention potato cultivation and consumption across different parts of the Old World from Ireland to Italy.[13] Yet he too reinforces the prevailing assumption that it was not until the late eighteenth and early nineteenth century that Europeans widely adopted the potato as a human staple rather than as animal fodder. This view is contested by historians like Rebecca Earle, James Walvin, or William H. McNeill, who find evidence of widespread potato farming for human consumption in Ireland, Britain, and the Low Countries during the second half of the seventeenth century.[14] Moreover, in East Asia, it was corn and sweet potatoes that first made significant inroads, starting from the 1560s in China's Yunnan province, followed later by their extensive adoption, together with the potato, in the Yangtze basin, Fujian, Szechuan, and Hunan during the second part of the eighteenth century.[15]

This staggered and patchy dissemination of the potato and its American sister crops, corn and the sweet potato, thus requires some explanation. Environmental historians like Jared Diamond and Alfred Crosby tend to emphasize the biological determinants of the process. For Diamond, as already discussed in the preceding chapter on the natural and the social, it is latitude that dictates the success of transfers: "Localities distributed east and west of each other at the same latitude share exactly the same day length and its seasonal variations. . . . Woe betide the plant whose genetic program is mismatched to the latitude of the field in which is planted!"[16] Similarly, Crosby underlines the climatic versatility of corn in explaining its global triumph: "maize will produce good crops in an extreme variety of climates. Its advantage over Old World plants is that it will prosper in areas too dry for rice and too wet for wheat. Geographically it fitted neatly between the two. Its supremely valuable characteristic is its high yield per unit of land which, on average, is double that of wheat."[17] Clearly, there are biochemical properties to plants and animals that condition their capacity to flourish in different environments. But as Marcy Norton and other critics of Crosby have suggested, absent in his account is the "social context, which largely determined what and how novel New World flora and fauna were appropriated."[18]

Consideration of the social context accompanying the Columbian exchange hence raises one of the axiomatic assumptions of modern social theory: namely, the constant interaction between human agency and environmental constraints in historical evolution—a tension between "the social" and "the natural" guiding this inquiry, and first broached in the previous chapter. From this perspective, the eastward migration of potatoes or corn and the arrival of wheat and cattle in the western hemisphere—and, even more so, the uptake of chocolate or tobacco in the Old World—are all a product of political and sociocultural structures and process, rather than determined biologically. Once again, environmental conditions plainly intervene in such interactions—as we'll shortly see, often causing unintended consequences like deadly epidemics. Habitat is, to be sure, one of several forces that combines with material and symbolic culture to generate particular socioeconomic and political outcomes. But *terroir* is not destiny.

Wheat, for instance, acquired such a powerful hold over the Mesoamerican landscape, not simply due to natural adaptation, but because it was the key ingredient in bread—a food item conquering Catholics believed was the body of Christ. Spaniards could and did eat corn-based starches like tamales, tortillas, or arepas. But social distinction (a term developed further in chapter 9) required that the latter, associated to the heathen worship of corn-protecting deities, be subordinated to the proper, Christian sacrament and symbol of cultural superiority—wheat bread.[19] Similarly, the diffusion of the potato across Eurasia was deeply conditioned by different degrees of state intervention and economic relations of exploitation. From the late eighteenth century, ruling classes from Britain to Russia, France, and Prussia promoted the potato as a cheap and filling foodstuff for their subaltern subjects, whether domestic or colonial, planting it extensively and legislating for its adoption by peasants, cottiers, serfs, and soldiers as their basic source of calories.[20]

The Columbian exchange, however, did not merely involve the natural transfer of flora and fauna across latitudes, it also entailed the movement of humans and, with them, their germs and diseases, as well as their cultures and cosmologies. Food became a central part of this story in at least two interrelated senses. First, American geography was in most places radically transformed with the arrival of Old World crops and ungulates. This was chiefly the result of Spanish settlers aiming to reproduce in conquered lands the patterns of consumption they had aspired to in the Iberian peninsula—plentiful bread, wine, fruit, lamb, and pork.[21] But as the colonial economy consolidated across Latin America, it was also integrated within wider imperial circuits of trade and commerce—thus, as we'll shortly see, skewing land use toward production of cash crops for export. Most of the continent's habitat turned into what Alfred Crosby labeled a "Neo-Europe": temperate regions of the world overwhelmingly populated by European settlers and their descendants, which subsequently became the planet's major breadbaskets and meat exporters.[22]

Mexico is a good example of this dramatic change, as extensive grazing and browsing livestock—much of it feral and quick to multiply—encroached on the existing, more intensive horticultural exploitation and its accompanying wild grasslands and shrublands. The Mexican highlands, most notably the upland Bajío valley north of Mexico City, in turn, became centers of wheat farming and export.[23] Summarizing a complex and protracted process, John F. Richards has suggested that "ungulates modify competition among plants species and, over time, change plant cover in grazed areas. The newcomers push back woodlands and scrub or woody vegetation and create mosaics of open spaces." Whether the latter are coterminous with ecological degradation remains a moot point among environmental historians of Mexico.

Some, like Elinor Melville, argue that the "ungulate irruption" was responsible for "irreversible damage to vegetation, social and water resources in Mexico," while other students of the same regions indicate that "high stocking ratios did not necessarily lead to overgrazing."[24] What does seem indisputable is that European conquest radically transformed the American landscape through a combination of "ecological imperialism" and the adaptation of New World land use to Old World patterns of food production and consumption.[25]

A second way in which food can be seen as a conduit and marker of the social change accompanying colonial conquest was the creation of distinct Creole diets. The Columbian exchange was in many ways felt most intensely in the epicenters of the encounter: the American and Caribbean crucibles where New World and Old World peoples, crops, animals, cooking techniques, and foodways combined in a syncretic fashion to deliver a new alimentary repertoire (not least through the skill and creativity of female concubines, domestic slaves, and servants of various ethnic backgrounds who cooked for European households). The *mole poblano* is often cited as a classic example of a Creole or mestizo dish in its blending of New World chocolate, chiles, and turkey with Old World spices, garlic, and onions.[26] Although, as Jeffrey Pilcher and Rachel Laudan forcefully argue, *mole* may be more the result of Creole elites wishing to emulate their Iberian peninsular counterparts than a novel, post-Columbian fusion food, there is no question here that distinctive regional cookery emerged across the American continent as a result of the admixture of eastern and western ingredients, recipes, and techniques.[27]

It is certainly important to note that, like most other societies, the new American food cultures were socially stratified and geographically differentiated—many poor Amerindians continued to eat the local staples of their ancestors, while white bread, wine, red meats, and spices were the preserve of an elite, not the entire Creole or mestizo population. Class, region, ethnicity, gender, and rural or urban location all played their part in conditioning the diet of the increasingly diverse American peoples. Indeed, as Africans and Asians arrived on American shores, forced or voluntarily, diasporic foodways made their way into the New World repertoire. A spiced spinach-based stew known as *callalou* in the Caribbean, *carraru* in Brazil, and (when cooked with okra) *gumbo* in the United States was carried over from west African culinary traditions.[28] So too was *fufu*, the west African glutinous dipping mass made from tubers, plantain, or taro that became known as *angú* in Brazil, *mangú* in the Dominican Republic, or simply *fufu de plátanos* ("banana fufu") in Cuba.[29] In later centuries, with the Caribbean arrival of indentured labor from South Asia, the goat or pumpkin roti became a local street food, combining elements of all three continents.[30] "The only way to become an American," Donna Gabaccia suggests of the colonial period, "at least as an eater was to eat creole—the multi-ethnic cuisine of a particular region."[31]

The point, then, is that the Columbian exchange occasioned the emergence of unique Creole or mestizo gastronomic cultures, a process captured in Fernando Ortíz's seminal concept of *transculturation:* a "transition from one culture" that does not, however, "consist merely in acquiring another culture, which is what the English word *acculturation* really implies, but the process [of] loss or uprooting of a previous culture, which could be defined as deculturation. In addition, it carries the idea of a consequent creation of new cultural phenomena which could be called neoculturation."[32] For this Cuban anthropologist, tobacco and sugar represent the wider contrapuntal culture emerging from colonial encounters like those that unfolded in the Caribbean. These two crops formed the material basis of plantation economies such as Cuba, where the dialectical unity of opposites—master and slave, white and black, technological rationalism and human brutalization—delivered a contradictory, conflictual, and hybrid society made up by the sum of the local and extraneous antagonisms: "Sugar cane and tobacco are all contrast. . . . The former seeks the light, the latter the shade; day and night, sun and moon. The former loves the rain that falls from the heavens; the latter the heat that comes from the earth. The sugar cane is ground for its juice; the tobacco leaves are dried to get rid of the sap. . . . Food and poison, waking and drowsing, energy and dream, delight of the flesh and delight of the spirit, sensuality and thought."[33] And yet both plants to this day thrive on the same island, both singular products of the Columbian exchange.

THE CONSEQUENCES

Among the most tragic consequences of the Columbian exchange—and there are several candidates, including Atlantic slavery, war, and environmental degradation—was the actual or near-destruction of aboriginal American populations through overwork, displacement, violence, and, above all, disease. Europeans and their animals brought to the Americas pathogens which infected un-immunized locals with diseases previously unknown in the New World like smallpox, influenza, typhus, measles, tuberculosis, and cholera. Lethal exposure to foreign microbes and viruses was compounded by a marked deterioration in the living conditions of the majority of Amerindians during the first decades of conquest. Dislocated through forced segregation into "Indian" reserves; super-exploited and undernourished in mines and fields; separated from their households and means of subsistence; traumatized by sociocultural oppression; and subject to vicious punishment and repression, Amerindians suffered a drastic decline in fertility and a disastrous rise in mortality that, by the mid-sixteenth century, had erased the indigenous presence in the Caribbean and reduced the mainland natives to about a tenth of their estimated preconquest numbers.[34]

A cruel irony of this "Great Dying" was that as Amerindians were decimated through contact with the outside world, the life expectancy of those inhabiting the rest of the globe increased, in large measure as a result of the foodstuffs like potatoes and corn imported from the Americas: initial depopulation of the New World contributed toward a gradual repopulation of the rest of the world, eventually including west Africa and the Americas themselves.[35] Following the original demographic collapse caused by the shock of conquest, the surviving autochthonous population slowly began to recover and—combined with the Old World newcomers—the Americas returned to demographic growth from the late seventeenth century onward.[36] Moreover, the incorporation of the Americas into the Eurasian food system contributed toward the "Rise of the West" during the early modern period (roughly 1500–1800) and the subsequent industrialization of that part of the world. The remainder of this section explores the long-term socioeconomic consequences of the Columbian exchange, focusing on debates around the role of the American conquest in the rise of industrial capitalism and the accompanying theories of how that encounter shaped the global food system we know today.

There is no doubt that the technologies, institutions, and profits associated with food production and trade across the Atlantic seaboard had a significant impact on the development of capitalism and industry first in Europe and then elsewhere. Sugar is perhaps the most emblematic food item in this regard as a mediator in the intimate connections between the plantation system (large-scale managed tropical agriculture) of the (circum-) Caribbean and capital accumulation and state-formation in western Europe.[37] The New World sugar plantations represented not only a large export market and a phenomenal source of wealth to be reinvested in European economies, particularly in Britain from the 1700s onward. They were also fueled by a slave trade (and its associated shipping, insurance, and finance sectors) that generated enormous profits and tax revenue repatriated to the mother country, organized through labor regimes that were in many ways replicated in European factories, and handsomely subsidized and underwritten by Old World banks and agencies that played a key function in providing credit, security, and liquidity to both metropolitan capitalists and colonial merchants, planters, and captains.[38] This pattern of unequal exchange across the Atlantic was reinforced and extended in subsequent centuries as Latin American and Caribbean economies, in particular, oriented much of their agrarian sector toward the export of primary food staples such as tropical fruits, coffee, cocoa, livestock, and grain.

In the course of the postwar decades, radical and Marxist theorists across the Americas began to analyze this division of labor on a world scale, defining it in terms of *structural dependency* between a metropolitan core of imperial states and their (formerly) colonial satellites.[39] In the phrase made famous by one of its most prominent exponents, Andre Gunder Frank, the economic development of the core was premised on the underdevelopment of the periphery.[40] On this account, the principal consequence of Columbus's discoveries was the subordinate incorporation of South America and other conquered continents into a capitalist world system, which relegated peripheral regions to the production of primary goods—food staples prominent among them—for processing, finishing, and marketing in the imperial metropole. While the capitalist core developed its economy and wealth on the back of this exchange, so the argument runs, the underdeveloped satellite societies were mired in the gross class inequalities, corrupt politics, poor infrastructure, and the persistence of slavery and forced labor that accompanied peripheral backwardness. Food and drink—in the case of circum-Caribbean rum, *cachaça,* or Southern hemisphere wine—became critical protagonists in this history of global capitalist development, insofar as they reproduced the plantation, ranch, and estate economies in the cultivation of peripheral cash crops and raising of cattle for sale in the metropolitan core.

Few today would question the contribution of food production and trade in the asymmetrical economic relations between the Old and New Worlds after 1492. Kenneth Pomeranz's careful analysis shows that access to the vast land resources of the New World, coupled with the enslaved labor that mined and farmed it, gave Europe a competitive advantage in its industrial "takeoff."[41] This is not, however, the same as arguing that the commercial networks that issued from the Columbian exchange were responsible for the advent of capitalism or, indeed, for the origins of industrialization, as discussed in chapter 5. Although it may appear a petty semantic distinction, there is an important conceptual difference between societies with markets (where exchange is embedded in all sorts of kinship, dynastic, or guild prerogatives and privileges) and a market society where economics is decoupled from politics and where virtually all members of society (be they bourgeois, proletarians, rentiers, or unemployed) are dependent on an impersonal cash-nexus for their reproduction.[42] If we apply this seemingly abstract distinction to actual historical societies at the center of the Columbian exchange and its aftermath, it is clear that Iberians didn't bring capitalism to the Americas, but rather grafted the feudal structures of the Reconquest raiding frontier to the tributary forms of exploitation they encountered among Inca and Mexica empires to create the unique *encomienda* and later *repartimiento* system of wealth appropriation. Similarly, the transfer of bullion and slave-trade or plantation profits to the Old World did not in itself transform Spain or Portugal into capitalist economies. In many respects, this long-distance exploitation of minerals and monoculture reinforced and prolonged the static patrimonial mercantilism that characterized these Iberian empires, leading to their eclipse by the bourgeoning capitalist states of Protestant Europe from the late seventeenth century onward. Yet, until well into the nineteenth century, even the British, French, and Dutch colonial empires continued to practice a mercantile form of capitalism where slavery, plunder, and privateering secured plenty and prestige abroad while industry, finance, and taxation flourished at home. Indeed, it was not until the second Industrial Revolution from the mid-nineteenth century onward that we can speak of capitalism as a dominant mode of production across Europe, let alone other parts of the world.

Meanwhile, societies most directly affected by the Columbian exchange across both sides of the Atlantic developed their own distinctive socioeconomic dynamics and political structures, which, in turn, had a momentous impact on the eating and drinking habits of their populations. The consolidation of colonial rule across the Americas occasioned the slow growth of domestic markets and regional commercial circuits. Sugar, rum, and molasses—as well as tobacco, indigo, spices, and the slaves who produced them—were trafficked between the Caribbean islands and the American Atlantic seaboard despite attempts by European powers to uphold their exclusive commercial monopolies. Indeed, piracy, smuggling, and contraband became especially pronounced across the region during the eighteenth century and beyond, as a means of furnishing local demand in foodstuffs and stimulants. Workers in mines and on plantations and estates also had to be fed, as did urban colonials, artisans, household slaves, and proletarians. Regional systems of food provision thus emerged so that Bajío wheat supplied both local and overseas markets, while the Andean potato, which the Spaniards had adopted as the food staple for Amerindian silver miners under their charge, migrated, during the sixteenth century, not just to Eurasia but also northward to Mexico for that same purpose.

To be sure, included in the food rations distributed at the time to slaves and other American and Caribbean workers—most notably sailors—were items traded across different parts of the colonial economy: salt cod from the Newfoundland banks; Irish dried or jerked beef; New England wheat biscuits (hard tack); rice from South Carolina; or local rum and imported beer and brandy.[43] But even (or especially) the most oppressive social regimes engender everyday forms of survival and resistance—food included—where resourcefulness, invention, and memory play an equal part. Testament to this was the sale of chickens, legumes, and fruits in local markets by slaves with access to small gardens that they cultivated during precious recreational time available outside the working week. For Sidney Mintz, "In slave societies, the ability to accumulate, like the ability to bequeath, becomes a symbol of freedom; where patrimony can lend dignity to genealogy, individual accumulation can mean individual identity. . . . Its growth attests to no instinctive desire for barter or for gain, but to the unquenchable human spirit of the slaves who nourished it."[44]

In the old continent as well, what until the late sixteen hundreds had principally been luxury markets in what Mintz labels "drug foods" (sugar, chocolate, tobacco, tea, and coffee) began to be democratized, in the course of following century, to the extent that on some calculations Britain's annual per capita sugar consumption shot up an astonishing 1,000 percent—from around two pounds in 1660 to twenty pounds in 1800.[45] Marcy Norton reports that in Madrid alone by 1722 there were 290 shops selling cacao and chocolate in a city of 127,000 souls—a chocolate retailer for every four hundred inhabitants![46] Across eighteenth-century urban Europe—and particularly in cities like London, Amsterdam, Hamburg, Bordeaux, Nantes, or Antwerp—an "industrious revolution" was underway. As Christopher Bayly neatly summarizes it, "the consumption of coffee and, later, tea went along with the purchase of sugar, fine breads, and replaceable plates off which to eat these items. The resulting package—let us call it 'breakfast'—gave people a higher calorific intake, a new time discipline and a new pattern for sociability and emulation in the household."[47] (To this day, neighborhood grocers in Spain selling items like sugar, chocolate, rum, or salt cod still carry the sign "Ultramarinos"—"Overseas"—above their shop front). The popularization of drug foods, most of them imported from the Americas, thus transformed not just consumption patterns across diverse social strata but also the European built environment itself, as the proliferation of coffeehouses, taverns, commercial wharves, stock exchanges, and warehouses, as well as seamen's lodging-houses, dockyards, and lascar quarters radically changed the topography and, indeed, the demography of Atlantic seaports.

There is no certain way of defining the "completion" of the Columbian exchange—clearly this was a process of long duration with multiple, uneven endings and legacies. Yet it is fair to say that in the course of the nineteenth century (the 1846 repeal of the British Corn Laws is usually cited as a key turning point), mercantilism slowly gave way to "free trade" and, with it, the widening world market in spices, drug foods, and enslaved humans created by the Columbian exchange was gradually deepened, but also replaced through the commodification of land and labor ushered in by the global spread of industrial capitalism. While the early-modern period of post-Columbian conquest witnessed a moment of "archaic globalization," the turn of the twentieth century belle époque marks the coming of age of a properly capitalist globalization. The Columbian exchange (literally) prepared the ground for, but was in most respects superseded by, global capitalism.

Harriet Friedman has written of a Settler-Colonial Food Regime, emerging under British hegemony, which from 1870 to 1914 "created the first price-governed market in an essential means of life."[48] This was a food system premised on the combination of nominally free wage labor and mechanized monoculture on extensive, industrially fertilized private land at the point of production (the "farm"), which, in turn, fed an increasingly internationalized, urbanized population with a disposable income, nourished mainly on a diet of sugar, cereals, meat, and alcohol at the point of consumption (the "fork"). The move from field to fork was, as chapter 5 on industrialization shows, accelerated by the growing rail, road, canal, and steamship networks, and the accompanying canning, freezing, pasteurizing, slaughtering, packing, and grain-storage technologies that broke down geographical limits of distance, climate, and season, as well as socioeconomic barriers to consumption of foods and drinks— from ice-cream to poultry—previously the preserve of the moneyed elite.

This chapter has linked the Columbian exchange to the era of European mercantilist empires—a period lasting until roughly 1815, when commercial exchange ("profit") was underwritten by all sorts of political-military resources and prerogatives of the imperial state ("power"). With the rise of nineteenth-century industrialization and the advance of national citizenship rights—what we'll see in chapter 5 described as modernity's "dual revolution"—a mercantile imperialism based on "buying cheap and selling dear" was replaced by a capitalist imperialism that largely succeeded in locking the world's working population and their land and natural resources into the value-producing "price-governed" or "self-regulating" market referred to earlier. A modern food system born from intercontinental exchange had now been transformed by industrial production into a properly capitalist food regime, which we shall be exploring at greater length in a later chapter on the political economy of the global food system.

THE LEGACY

To finish, picture a nocturnal scene in Madrid at the beginning of the twenty-first century: a group of friends mark the end of a night out at the Chocolatería San Ginés, just off the capital city's Plaza Mayor. A hot chocolate drink is prepared and served, together with the requisite deep-fried, sugarcoated dough sticks *churros* by an Ecuadorian national, one of over a million Latin Americans who in the 1990s migrated to work in Spain's booming economy. This type of contemporary social interaction is one product of the Columbian exchange. In fact, as we've seen in this chapter, it would not have been out of place in Madrid some four hundred years ago—with one important exception: very few Latin Americans, let alone Ecuadorians, would have been able, possibly even willing, to freely settle in Spain before the late twentieth century. The mass migration of Europeans to the Americas in the centuries after 1492 has only been reciprocated more recently, arguably resulting from processes associated to capitalist globalization. Attention to the spatial components of *geographical exchange* thus forces us to think about the temporal dimensions of *historical change* that accompanied it: what is the balance between continuity and rupture in the Columbian exchange's legacy when thinking about food, politics, and society today?

One way of recognizing the unique contribution of the Columbian exchange in contemporary gastronomies is to focus on the postcolonial combination of food styles and ingredients. It is no longer just in global cities like London, Tokyo, Paris, Sydney, and New York that all sorts of ethnic and fusion foods and international beverages are available at different price ranges. Urban consumers in many emerging economies—from Turkey to South Africa, Brazil, and Thailand—have for several decades had access to such global culinary offers in restaurants, bars, cafes, and supermarkets. The globalization of the catering industry has, moreover, facilitated the circulation of chefs, *stagiaires*, waiters, and sommeliers across continents, sharing cooking and serving techniques, recipes, skills, and knowledge that migrate with the staff themselves. Indeed, one notable phenomenon of the last thirty years or so is the rise of celebrity chefs (as explored in chapter 12) and cookery writers hailing from or representing the Global South across the metropolitan North.[49] It is reasonable, therefore, to suggest that the last five hundred years have witnessed a veritable "creolization" of food cultures throughout different parts of the planet and across different social strata. Of course, if we think of fusion cuisines as the confluence of various ethnic and regional cooking traditions, then plainly frontier regions and imperial capitals have for millennia delivered instances of multicultural gastronomy—but the distinctiveness of the post-Columbian age, as suggested earlier, lies in its global scale and the deepening of these consumption patterns within and across societies.

Yet this celebration of globalized consumption needs to be tempered with a close look at the relations of production that sustain these new transnational gastronomies. The market integration of the global food system initiated by the Columbian exchange has been accompanied by the continued social stratification of that very same food chain through class, nationality, gender, and ethnicity. Although slavery, mercantilism, and tributary exploitation are mainly a thing of the past, their legacies—in the form of poverty wages, rich-state protectionism, and resource exhaustion—are very much alive. They are still unevenly distributed across the world, with an overrepresentation of peoples from the Global South among the low-paid, unprotected workers—small-scale farmers, fruit pickers, truck drivers, meat packers, kitchen porters, restaurant waiters, toilet cleaners, and waste collectors—at every stage of the food cycle. Throughout the advanced capitalist world, the promise of gastronomic cosmopolitanism has been compromised by immigration controls that choke the supply of overseas kitchen and serving staff.[50] In the United Kingdom alone, labor shortages are reportedly threatening the survival of a sector made up of some twelve thousand curry houses employing an estimated hundred thousand workers and generating

£4.1 billion for the British economy.[51] The precarious livelihoods of food workers (from farm to fork and beyond) thus continue to be a structural condition for the reproduction of global, hybridized cuisines.

As we shall see in chapter 10, "Political Economy," this latter reality has led some to argue that for billions across the world, food choices remain heavily constrained by neocolonial structures of uneven trade and commodity exchange. A politically independent Senegal is, for instance, still significantly dependent on rice imports from Vietnam, as under French colonial rule local staples like sorghum and millet were displaced by ground nut plantations while imperial authorities flooded west African markets with rice cultivated in French Indochina.[52] During the period of postwar decolonization, the United States used food and other aid, famously through the Marshall Plan and the 1954 Public Law 480, to reorient its allies' food systems toward wheat and livestock.[53] On this reading, then, the continuities with the colonial mercantilism that issued from the Columbian exchange far outweigh the ruptures, and we must still speak of an imperialist global agribusiness sector.

The view adopted here has been different. Although there are plainly abiding structural inequalities to the global food system, these are principally generated by the globalizing tendencies of capitalism as a mode of production, rather than by imperialist attempts at controlling the circulation of commodities. The merit of this distinction is, once more, to focus attention on the logic and dynamics of the "price-governed" or "self-regulating" market that, though aided and abetted by state authority and multilateral organizations, is mainly responsible for the socially unequal and geographically uneven distribution of food in the world. Such emphasis on moments of radical historical transformation—be it the conquest of the Americas or the advent of industrial capitalism—guards against an infinite regression where all experiences of long-distance carrying trade or any example of intercultural admixture of food and drink become markers of a perennial globalization or hybridization of the world. As this chapter has sought to illustrate, while these latter phenomena are by no means unprecedented, they have found a very distinctive socioeconomic and political expression over the past few centuries, and the Columbian exchange lies at their origin.

What has also hopefully been conveyed here is the power of various concepts and theories in enriching our understanding of the globalization of food, politics, and society in the modern period. Substantive interpretations of exchange, like those of Sidney Mintz or Eric Wolf, among others, encourage us to think about the social and political content of seemingly natural and self-regulating economic transactions mediated through markets. Alfred Crosby's seminal study and those of other environmental historians, for their part, force us to consider humanity's collective metabolism with nature: the role and significance of biophysical forces upon the structure of human societies, and vice versa. Symbolic and material culture, as theorized in Fernando Ortiz's notion of transculturation or in various conceptions of creolization and *mestizaje,* have also been shown to play a critical role in explaining the unique foodways of modern America and beyond. Finally, the work of political economists and world historians following various methodological traditions (though many of them inspired by Fernand Braudel's framework of "total history")[54] bring to the fore crucial questions of periodization and change in human evolution. Beyond telling us how the chili got to South Asia and the lime to South America, these stories of food transfers disclose the powerful sociocultural processes and political structures that underpin not only the inequality, oppression, and unevenness but also the diversity, innovation, and abundance that characterize the global food system today.

CHAPTER 4

Culture

Ritual, Prohibition, and Taboo

When Helen Sharman, the first British astronaut, arrived on the Mir space station in 1991, she was greeted by the cosmonauts with a chunk of bread with a salt tablet stuck into it.[1] Given that they had to make do with the available prepackaged ingredients of the space larder, this enactment of the traditional Russian welcome for respected guests may have lacked some of the gastronomic charm of its earthly original—usually a whole, decorated loaf of bread with a bowl of salt at its center, and the guests and their hosts break off pieces of the bread and dip them into the salt—but the quality and exact form of the food was not the point. In this ritual, common across many Slavic, Germanic, and Arabic cultures, as well as in the Russian Orthodox church, the bread and salt stand for very much more than simply themselves, and more, even, than their basic symbolism—respect (in bread) and friendship (in salt)—or the components of the Russian term for hospitality, *khleb-sol* (literally "bread-salt").[2] The ostensibly simple performance masks a complex network of social and cultural signifiers that go far beyond the food. We could say that in this instance, Sharman's new colleagues were offering her the proper welcome into their culture, acknowledging her status as one of the team, emphasizing the Russianness of Mir, and normalizing the extraordinary experience of being in space by connecting with the social mores of Earth. We could interpret Sharman's recognition of and participation in their ritual as signals of her understanding and accept-ance of her hosts' culture, her willingness and ability to become socialized within it, and her respect for the social rules of the new microcosm of society she had traveled to.

There is no need to travel into outer space to find examples of these kinds of "traditional" food practices, but it perhaps tells us something about the representational, symbolic power of food that humans tend to invent, reinvent, and perpetuate such rituals wherever they go. Food, in its broadest sense, is a universal cultural object—a "primitive physiological fact," as identified in the introduction—and as such, it often functions as a powerful expression of our social and sociopolitical organization. Rituals reinforce collective representations: they express something about the conception the group has of itself in relation to the world, both its place on the planet and in relation to the rest of humankind. Ritual also implies repetition and habit, and thus history, whether real or imagined. This chapter explores food as this contemporary and historical representation of social, cultural, political, transactional, and religious systems, and it examines the possible combinations of these as sources of agreement or conflict. It does so by looking, in turn, at various themes that draw out one of the core propositions of the book concerning the particular relation between self and other under modern conditions. It discusses prohibition and rules of conduct around food and eating: the what, when, why, who, and how of various cultural practices. It considers the significance of these practices, elaborating on them as a reflection of social structures, social relations, kinship, economy, and politics that various modern anthropologists, in particular, have explored using notions of purity, symbolism, myth, and danger. It investigates some of the ways in which food provides the material that determines and regulates social relations; how those systems, symbols, and categories developed; and what they mean according to historical anthropological, ethnographical, and sociological studies. For when incorporated into the practices of any culture, food both reflects and reveals societies' fundamental ideas, whether about power relations, gender roles,[3] or the boundaries between people, animals, gods, and nature—what is "good" or "bad," what is permitted or not.[4]

In defining boundaries and making distinctions, elements (in this case, foodstuffs) that fall outside them may become taboo (literally "set apart" in the original Polynesian conception of the term), forbidden, or prohibited; and those people or groups that follow different rules, who practice customs (or in this case, eat) what we do not, become definitively "other." While the drivers for these self-definitions may spring from various social or cultural sources, they are often either part of long-standing myth or codified as rules within religions or other social groupings. As Frederick Simoons pointed out in his work on different meat-eating prohibitions around the world, past and present, food practices are a common means of group identification, a means of recognizing and emphasizing similarities and differences. He explained that the Tibetan people of the Kansu-Tibetan border region explicitly identify food habits with other religions and cultures, apparently asking of other groups: "Is their mouth the same as ours, or is it like the mouth of the Moslems, or do they have some other mouth?"[5] To consider this question of the significance of other people's "mouth," and what it might mean, it is informative to examine religious prescriptions around the selection, preparation, and consumption of food with a focus on rules of prohibition, such as laws of ritual purity in Judaism *(kashrut, kosher)* and Islam *(halal)*. In doing so, we address notions of the sacred and the profane, as well as uses of fasting and feasting in religious observance.

ANTHROPOLOGICAL ANALYSES

Anthropologists, especially social anthropologists, have been highly influential in developing a theoretical basis for studying food in culture and society, and much of the twentieth-century work that remains important emerged out of late nineteenth- and early twentieth-century ethnographic studies that considered food for its symbolic value in kinship, religion, and sexuality. Despite a wide variation in methods and conclusions, the new discipline of anthropology shared with other modern social sciences that accompanied global industrialization (sociology, geography, psychology) an Enlightenment focus on the human, as opposed to otherworldly, explanations for the complex relationship between food, customs, and cosmologies. The approaches and assumptions underlying these early studies have been subject to extensive reexamination and critique. However, their influence on literature and the history of ideas in particular makes an acquaintance with their concepts helpful in developing an understanding of current approaches. Moreover, they underline the value of an anthropological study of food when considering some of the core binaries present across this book—the creative tensions in the relationship between nature and society, self and other, and, indeed, our public and private lives. In our more general examination of the role of "culture" in a social theory of food, a brief critical historiography of some of the key writers in the earlier periods provides some context in which to reconsider the framing of today's debates as entirely contemporary.

James George Frazer's *The Golden Bough* (1890), an early and much criticized anthropological work, discussed belief systems in societies around the world, proposing a model of social and cultural progress from the "primitive," where magic prevailed to more sophisticated structured religions, which, in turn, would be replaced by science.[6] Focusing on the relationship between cult, myth, and ritual and the consistencies between origin stories across divergent cultures, his work shared common ground with William Robertson Smith's examination of the similarities between early Hebrew folklore and the Old Testament, as well as his use of sociological techniques to analyze religion.[7] Their sociological approach meant that both writers highlighted the importance of collective eating, or commensality, for group solidarity as well as for its place in other ritualized behaviors, particularly religious behaviors. Frazer emphasized the importance of totems in what he called the "primitive" stage of development, which included animist and naturist religion.[8] Contemporary scholars, of course, no longer measure other cultures, whether in different periods of history or different parts of the world, on a scale from primitive to civilized, but the idea of "progress" embedded within the narrative of modernism means that we do still come across the uncomfortable and archaic implication that others—people with different habits and practices than ours—are at an earlier stage of development towards true "civilization" (or being more like us).

In Frazer's work, the other was defined by a lack of perceived sophistication, characterized particularly by proximity to nature rather than culture, and this could be readily represented by the group's identification with totems. Totemic objects are usually derived from nature, being either plants or animals, and are subject to clear and stringent rules within the group that adopts them. In general, totems are not eaten. For example, the northern tribes of central Australia are responsible for increasing the living numbers of their totemic animals and thus banned from eating them.[9] However, as sociologist Émile Durkheim pointed out, the status of totems as "holy things" means that within other cultural structures they can also "enter into the composition of certain mystic meals" or even serve as sacrifices or sacraments.[10] Ritual consumption of the holy object by the deity (in an offering) or the members of the religion (in a ceremonial fashion) both affirms the status of the totem and defines membership of the group.

Sociological work moved beyond the consideration of totems as objects of superstition to emphasize the social meaning of food customs or myths. In their essay *Primitive Classification* (1903), Marcel Mauss and Émile Durkheim suggested that the spatial organization of the world reflects social organization and that society provides the categories of human thought.[11] Conceptions of time and space are socially constructed, and we are divided and differentiated accordingly—ideas explored in more detail in Durkheim's *Elementary Forms of Religious Life* (1912), which discussed the role of collective representations in the categorization of nature. In the early 1920s Bronislaw Malinowksi (1884–1942) extended this thinking by developing a functionalist approach to ethnographic study of social organization within cultures. He suggested that all human institutions are designed to satisfy fundamental human needs (including our common need for food), meaning that our cultures share more than we might think. His detailed study of the Trobriand people is a pioneering demonstration of "participant observation," a technique in which the researcher conducts his or her research from the inside, by taking part in the daily life of the research subjects and experiencing their culture firsthand.[12] This technique is now common practice in anthropology as well as sociology, and in food studies, it has been particularly valuable, as it allows for a detailed examination of food in all its facets, including as a social system in itself and a primary mode of social organization.

THE CULINARY TRIANGLE

Claude Lévi-Strauss's (1908–2009) writing, in particular the *Mythologiques (Mythologies)* volumes *The Raw and the Cooked* and *The Origin of Table Manners,* effectively rewrote the earlier approaches to myth. He was extremely influential in establishing the structuralist idea that food is a language and cultural system in itself.[13] In his model, every aspect of food becomes a patterned activity replete with symbols and with its own structures of meaning, whether directed at ideas about what is "natural," "virtuous," "appropriate," "civilized," or whatever else might be projected by or onto this cultural artifact.[14] In this context, he developed a model he called the "culinary triangle," a schema that continues to be referred to in food studies, in particular as a depiction of the relationship between nature and culture and of the universality across cultures of the distinctions between them.[15] Although it is ostensibly simple, particularly in its earliest versions (before the inclusion of

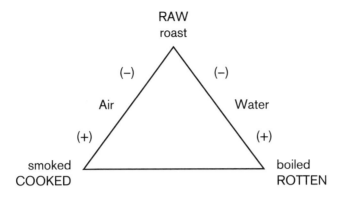

FIGURE 2. The Culinary Triangle. Claude Lévi-Strauss.

cooking method), it can be confusing and easily misunderstood, so we'll pause to discuss it here.

First, it is worth unpacking the name of the model: it is a *culinary* triangle. In other words, it is a paradigm to be applied to groups with a concept of cuisine in Rachel Laudan's sense of food as "something that humans *make* in the kitchen and the processing plant."[16] The word gives us the first hint that here apparently straightforward, readily understood food words are being applied to some quite slippery theoretical ideas. This is not, as it might appear, simply a way of categorizing the essential nature of foodstuffs, and it does not necessarily refer literally to cooking methods. As much as there might be literal examples that could slot into each category (such as a salad composed of raw ingredients), at the same time, the terms are also used conceptually. Which foods or dishes fit into which category shifts depending on who is thinking about it and what their concept of what constitutes a cuisine or the culinary is.

The key words—*raw, rotten, cooked*—are translated from the French, *cru, pourri, cuit*. Although these are correct translations, and perhaps the most obvious and compelling choices of word for a culinary model, it is worth noting that there are other possible options. The original French words contain within them layers of meaning that are somewhat lost in the more specific English. A consideration of some of these nuances of meaning is particularly helpful when we come to the words relating to cooking method that are attached to each one of the main concepts, at each corner of the triangle. How can something "raw" also be "roast"? Who would eat something "rotten" even if it had been "boiled"? And why is the "cooked" specifically "smoked"? This begins to make sense if we can think beyond the ideas of edibility embedded in the English words and take on the cultural implications that also reside within the French ones. To do this, the elements shown to be taking action down either side of the triangle—air on the left, water on the right—as well as the fire implied at the upper apex, provide helpful pointers. The culinary triangle assumes a more sophisticated mode of eating and food preparation than simply hunting and gathering: in particular, that there is some form of intervention beyond, say, plucking a berry from a bush.

"Raw" *(cru)*, at the pinnacle of the triangle, might be more helpfully translated as "crude" or "simple," things that are more "primitive" or basic. This category refers to the emotional or instinctive, to foods that are the least acted upon by external factors and hence are in their least processed (or socialized) state. The model assumes that fire is available for cooking, hence the most basic form in the model, the raw, is described as "roast." This makes sense when we consider that although humans have developed many tools and techniques for cooking with fire, it is possible to roast meat using no cultural artifacts at all, simply by placing it in or next to the embers or flame. If we were to take an example of a foodstuff like bread, seen from the perspective of a contemporary foodie, perhaps a member of the Slow Food movement[17] or the Real Bread Campaign,[18] we might put an unleavened flatbread or ash cake into this category—flour and water, mixed, shaped, and placed directly onto the flame or into the embers, producing a fast, simple, and serviceable basic bread.

"Rotten" *(pourri)*, though it could arguably apply to fermented foods in general, should not necessarily be thought of in the sense of something having gone off and become inedible. Rather, these are foods that have been partially processed and are thus rotten in the sense of "corrupted" or acted upon by external factors. Lévi-Strauss chooses water as the element that expresses this transition, explaining that this medium allows for foods prepared with a greater degree of sophistication than with the simple application of fire. In order to use water in cooking, one needs to find or create a cultural artifact, a vessel to hold it. In simple terms, this might be a stew of several ingredients placed in a pot over a heat source. These "rotten" foods could be seen as the "raw" in an intermediate stage at which they have been slightly "cooked" either by natural processes (as in the fermentation and brining of olives, where no heat is applied but time is invested to see the process through) or by relatively simple cultural processes (as in boiling some ingredients in a pot with liquid). For a subtler interpretation, we can turn again to our contemporary bread example. An industrially produced white loaf, a yeasted fermentation of flour and water baked in a steam oven, thought of as the definition of bread by vast swathes of the world's population, might fit into the "pourri" category as far as our Real Bread aficionados are concerned. For them, the kind of processing it has been subject to (especially the speeded up "Chorleywood Process" introduced in the United Kingdom in the 1970s) makes it both literally and figuratively "rotten."[19] What is food for many, is inedible to others: a non-food food. In this case, it is easy to see that the application of the category to the foodstuff is cultural, not definitive.

In turn, "cooked" *(cuit)* represents food that is fully processed or socialized and that tends to be bound by rules and conventions. It might be helpful to think of this as a state of being "ready" or "done" in the sense of a well-done *(bien cuit)* steak. Lévi-Strauss characterizes these cooked foods as having been smoked, or acted upon by the air, in the sense of their having been acted upon over a longer period of time, using a combination of nature (the smoke from the fire) and culture (the apparatus needed to place foodstuffs in the smoke). Returning to our Real Bread campaigners, for them an artisanal sourdough loaf—understood as a dough crafted through a long, slow process of harnessing wild yeasts and naturally developing lactic fermentation, a degree of hand-forming to the loaf, and perhaps a baking in a wood-fired oven rather than an industrial steam oven—might fall into this category.[20] Of course, for the majority of people whose daily bread is the soft, white industrial loaf, their version is the "cooked," and this kind of heavier, chewier, less technologically designed artisanal loaf might actually fall into the "rotten" category. As with all cultural interpretations, it all depends on the teller of the tale.

Roland Barthes (1915–1980) built on Lévi-Strauss's notions that our thinking about food is grounded in myth or the stories we tell ourselves as he laid out the semiological prospects of "the food [signifying] system."[21] Barthes saw food as a series of cultural signals, outside language, to be "read" as part of an investigation into the deep cultural roots and complexities inherent in everyday objects and experiences.[22] His explication in *Mythologies* (1957) of the "plastic powers" of wine (and "totemic" foods in general), along with his assertion such objects "can serve as an alibi to dream as well as reality, it depends on the users of the myth," allows the food scholar to apply this form of analysis to thinking about any food system, whether structured by religious rules (such as the use of bread and wine in Christian communion) or some other codification such as advertising.[23]

These kinds of approaches have, of course, been challenged. Although she later moderated her criticism, Mary Douglas (1921–2007) suggested that Lévi-Strauss was "orbiting in rarefied space" in his attempt to find universal meanings in food, and she said that his arguments were too closely allied with "unverifiable" evidence, particularly that found in literature.[24] Nonetheless, in her own highly influential work, structured observation leads to a similar conclusion—that food is revelatory of both implicit and explicit meaning in social relations that may vary in application but spring from consistent motivations. Douglas's anthropological examination of numerous aspects of specific cultures, including healing, taboo, purity, and pollution, supplies great insight into the interpretation of current and past societies, and it provides a model for the part food might play in such analyses. In her essays "Deciphering a Meal" and "The Abominations of Leviticus," she considers what the "precoded message" in food might be, in terms of the family and social and religious rules.[25] Douglas's review of the latter work, written decades later, suggests that one of her important findings was that the prohibitions in Leviticus (which we will look at in detail later in the chapter) are actually designed to protect the rest of the planet from human predation, supporting the broader biblical traditions that "always present[s] God as loving and caring for his creation."[26] Her continuing analysis of the basis of her own interpretation demonstrates just how fluid a cultural analysis can be. It depends not only on "objective" examination of facts, but also subjective interpretations that may vary according to the shifting perspectives of time, place, and even fashion. Although the symbolic role of food in culture seems undeniable, symbols in themselves are not universal or necessarily enduring, and myths might be interpreted in several ways.

RELIGIOUS RULES

This flexibility of interpretation applies even in instances that may appear on the surface to be more absolute in nature, such as religious definitions of the holy or the sacred and related practices. Durkheim explained that "a religion is a unified system of beliefs and practices relative to sacred things, that is to say, things set apart and surrounded by prohibitions—beliefs and practices that unite its adherents in a single moral community."[27] Yet it is obvious that there are numerous differences not only between religions but also between the various schools of thought within specific religions. Over centuries, differences between various Christian churches have been expressed in their approach to the practice of Communion. The question of whether the sacred consumption of bread (the body of Christ) and wine (the blood of Christ) is a literal or symbolic act, and to what degree, has divided Catholics from Protestants and different Protestant groups from one another since the Protestant Reformation in 1520, when Martin Luther described transubstantiation as "a monstrous word for a monstrous idea."[28] The Catholic belief that the bread and wine literally become the body and blood of Christ, reaffirmed at the Council of Trent in 1551, contrasts with the symbolic nature of Jesus's presence at the sacrament emphasized by Lutherans, Methodists, and in the thirty-nine articles of the Church of England (since 1571).[29] For Baptists, the bread and wine is purely symbolic of the Last Supper and is not thought of as a sacrament, but rather as an ordinance. While on the surface, the components of the ceremony may seem to be the same for all Christians (apart from Quakers, who do not practice a bread- and wine-based ceremony at all), the beliefs underlying the ceremony act as forms of differentiation, as well as a means of othering those with a different approach.

JEWISH KASHRUT

The Tanakh, or "written Torah"[30] (the books known to Christians as the Bible's Old Testament), contain the laws by which God expects humans to live. This includes a series of rules known as the kashrut, found in the eleventh book of Leviticus and the fourteenth book of Deuteronomy, as well as some passages of Exodus.[31] Instructions on which foods may and may not be consumed and how they are to be prepared and eaten are given; food that falls within these parameters is known as *kasher* or *kosher.*[32] The verses declare that certain animals, fish, and birds are "pure" and may be consumed, while others are "impure" and may not: "Of their flesh shall ye not eat, and their carcass shall ye not touch; they are unclean to you."[33] Edible land mammals must have divided feet, cloven hooves, and chew their cud in order to be pure, a trio of rules that allows the consumption of sheep, goats, and cows but excludes nonruminant pigs as well as rodents and camels.[34] Specific birds are forbidden. No general rule is offered to help categorize unnamed fowl, but all of those named as impure are either birds of prey like eagles or carrion-eating scavengers like crows.[35] Fish with fins and scales (like tuna, haddock, or cod) are permitted; those without (like sharks, eels, or monkfish) are not.[36] Nor are water creatures that are not fish, such as shellfish and mollusks. Slithering creatures like snakes and insects—with the exception of locusts—are not to be eaten.[37] Blood is always impure, since it is said to contain the very life essence of the animal, requiring that creatures should be bled and specifically bloody parts, such as the sciatic nerve, should be removed and disposed of during slaughter.[38] It is also forbidden to consume meat and milk together, a proscription that extends from the kitchen, where the two foods may not be cooked together in the same dish, presented on the same plate, or prepared using the same utensils, to the dining table, where meat and milk products may not be served at the same meal or even within several hours of one another (so that they cannot mix in the stomach).[39]

PURITY AND HOLINESS

Orthodox believers are expected to obey the rules, which, with the exception of the blood prohibition are simply laid down and not explained in the Torah. Some rabbinical texts, written to guide followers of the faith, elaborate on the instructions, emphasizing the importance to the Jewish faith of recognizing and respecting distinctions. The rules around food could be seen as an everyday manifestation of a practicing Jew's ability to distinguish between things that are defiled, evil, or wrong and those that are pure, good, and right. Obeying the strictures requires the exercise of self-control and demonstrates a broader ability to behave appropriately. They are a call to daily holiness, a regular reminder to respect the sacred over the profane and to effectively treat every act of eating as a religious ritual.[40]

Nonreligious scholars too have given thought to the reasons for the existence of kosher and other religious dietary laws, and many different explanations have been offered for them. For a summary of the most frequently presented of these arguments, put forward in various combinations by numerous scholars at different times, see table 1.

Mary Douglas analyses and rejects most of these arguments in her influential work *Purity and Danger* (1966), in which she assesses the Jewish religious food laws in the same way as the rituals and taboos governing other societies.[41] Her conclusion is that, like other cultural prohibitions, these ones are designed to emphasize to members of the society that "this is a universe in which men prosper by conforming to holiness and perish

TABLE I   SUMMARY OF PROPOSED LOGICAL BASES FOR BIBLICAL FOOD AND DRINK PROHIBITIONS[a]

Argument	Counterargument
The rules are ethical and disciplinary rather than symbolic.	This interpretation can be attributed to Hellenistic influence on early Jewish culture, but there is no clear argument against the rules containing at least some element of symbolism.
The rules are based on "natural" reasons, including health and sanitary concerns. This applies in particular to pigs, which are impure in habit (omnivorous, dirty) and whose meat might case the disease trichinosis.	There are no entirely convincing supporting arguments. The concerns about the uncleanliness and carnivorousness of pigs could apply equally to some animals that are permitted (e.g., chickens), so if God was banning pork on hygiene grounds, why not also ban these others? Trichinosis was not known, and other poorly cooked meats can equally cause illness.
The rules are designed to proscribe the most delicious meats and reduce the temptations of gluttony (also a health argument).	An argument put forward most prominently by Philo, but being a matter of taste, very unconvincing.
The rules, especially against pigs, are based on the appropriate farming practices for the regional ecology.	Other local tribal groups had no issue with raising, keeping, and eating pigs.
The rules are arbitrary and irrational or rigidly legalistic and cannot be explained.	This is unlikely: legislators tend not to legislate nonsense, and Leviticus is a priestly source concerned with order.
The rules are allegorical (i.e., they are not meant literally but outline the differences between virtue and vice).	Most opinions along these lines are pious commentaries rather than interpretations; they tend to be inconsistent and not comprehensive.
The rules are designed to protect the people from outside, foreign, or heathen influences and/or demonic creatures.	This is not consistently applied. For example, sacrifice, a long-standing foreign and historical practice, is retained.

NOTE: Compiled by Jane Levi

[a]Multiple sources, notably Mary Douglas, *Purity and Danger* (London: Routledge, 2002); F. J. Simoons, *Eat Not This Flesh* (Madison: University of Wisconsin Press, 1961); Marvin Harris, *Good to Eat* (London: Allen & Unwin, 1986); and Marvin Harris, "The Abominable Pig," in *Food and Culture,* ed. Counihan and Van Esterik (New York: Routledge, 2008).

when they deviate from it."[42] This idea of holiness is an important one, as it is, in many ways, synonymous with notions of completeness, perfection, and cleanliness. The kashrut is an example of an orderly system that "requires that different classes of things shall not be confused,"[43] so that, as Jean Soler points out, the basic requirement of any creature permitted for human consumption is that it should be appropriate to its natural element as defined in God's acts of Creation.[44] Following Douglas, Soler suggests that "in order to be considered pure, an animal must move,"[45] and that its means of locomotion are significant to its degree of purity. Thus, land animals should move across the earth with legs, not slither or creep on the ground;[46] fish should have scales and fins for swimming, not shells like mollusks or legs like crustaceans; and birds should fly.

HALAL

Qur'anic prohibitions are often characterized in relation to those of Leviticus, and since the Qur'an is a later text, written in the seventh century C.E., it is reasonable to assume some degree of influence. However, despite apparent similarities, for believers there are important distinctions between halal and kosher.[47] An observer of kosher rules may not substitute halal meat, since it has not been subject to the appropriate blessing; equally, kosher is not a substitute for halal. Having said that, there are variations in practice between different Islamic schools, just as there are major differences in daily practice between Orthodox, Conservative, and other observers of Jewish rules. For example, some Muslim authorities have ruled that a Muslim may eat kosher food if no halal is available, the requirement to respect and preserve human life taking precedence over other considerations.

The basic system of prohibitions is common across all Islamic sects. Blood, carrion, and swine are forbidden, as is the consumption of any food not consecrated to God, including meat that has not been ritually slaughtered in the name of God (Allah).[48] This latter prohibition means that although Muslims could ostensibly consume meat slaughtered in accordance with the kashrut, in practice it is forbidden—other than in the case when starvation would be the only alternative[49]—since the proper ceremony has not be been followed. Any foods that fall into these categories are defined as *haram* (forbidden, unlawful) or harmful and subject to definitive prohibition. In addition to pigs, there are other animals not named in the Qur'an that have been explicitly forbidden in some jurists' interpretations of the text and described as *makruh* (detestable), which are subject to advisory prohibition. These include predators, dogs, snakes, crocodiles, weasels, pelicans, otters, foxes, elephants, ravens, and insects (with the exception, as in Leviticus, of locusts). There is a different approach to sea creatures to that in the kashrut. Since the sea is considered to be pure in essence, all marine animals, including mollusks and shellfish, are halal according to most authorities. This applies even if they died spontaneously (while land animals in that state would be considered carrion). However, just as in the Talmudic interpretations of the written Torah, there are variations in the implementation of the Qur'anic instructions, seen clearly in the case of sea creatures. Shi'ite Muslims prohibit the consumption of fish without fins and scales, as do some Sunni Muslims (following the Hanafi school of law).[50] Prawns, however, are permitted to some Shi'a, depending on the authority consulted; some of them regard prawns' shells as a form of scales, and others consider them to be the water-based equivalent of locusts.

Alcohol and other stimulants are forbidden, but in a different context to foods.[51] Whereas foodstuffs are defined as being inherently clean or not, stimulants are prohibited on grounds of their impact on social order. They also provide a tool for social distinction between Muslims and others, as well as between different groups of Muslims. Where permissiveness on this matter arises in particular groups, in particular times or places, it is not in the context of canonical approval—the Ottoman Janissaries, who consumed alcohol, coffee, and tobacco, were members of the Bektashi order, largely considered to be heretical. Other consumers of these substances were wealthy social elites considered able to manage their own private behavior. Murat IV, sultan of the Ottoman Empire (1623–40), prohibited wine, coffee, and tobacco to his subjects, on pain of death, though he was himself a well-known drinker.[52] He was said to prowl the streets and waterways of Istanbul in disguise to monitor his subjects' conduct. On one boat trip across the Bosporus, the story goes, he was enraged to discover that the bottle the boatman was swigging from (and shared with him) contained wine, and he furiously condemned him for his disobedience. "Who are you to threaten me?" asked the boatman. "I am the sultan!" he shouted. "Hah!" the boatman replied mockingly. "One swig, and you think you're the sultan!"[53]

The distinctions between what is clean and what is polluted (not clean), what is permitted or not, can be expressed in numerous ways that may or may not be codified within a religion but are always cultural. As we have seen with both Jewish and Islamic practices, the rules are a way of separating foods into their appropriate and inappropriate kinds, and this notion of separation extends beyond the food itself to the people consuming it. According to one commentator on halal adherence, the Hui Muslims in China saw halal eating as a way of differentiating themselves from Han Chinese "values and traits."[54] The nonconsumer differentiates the fundamental character of the people through their food. If pigs and pork are dirty and insanitary, polluted in substance and by nature, then so are those who consume them. As Douglas points out, "the only way in which pollution ideas make sense is in reference to a total structure of thought whose keystone, boundaries, margins and internal lines are held in relation by rituals of separation."[55] Culture separates us into groups that either do or do not eat particular things, and these choices rarely have much to do with whether or not something is technically "edible."

DISGUST AND THE OTHER

At this point, it perhaps makes sense to take a short digression into notions of disgust, a natural companion of many food-based rituals of separation or othering, whether as cause or effect. When it comes to food, disgust is one of the most powerful emotional responses humans might have. As much a cultural response as an instinctive sensory one, it can be based on personal distaste (I don't like it), a perception of danger (I think it's rotten), or notions of inappropriateness (I/we don't think of that as food).[56] This complex set of responses, shared by all humans, has been described by researchers as "core disgust." A direct association between food and the self, implied by the oft-quoted truism that we are what we eat, is at the heart of the idea of core disgust.

It has been argued that disgust evolved as humans evolved and that its development is particularly linked to the Neolithic period. As discussed in chapter 2, this is when humans began to herd and domesticate animals, more regularly consuming meat and dairy products and developing ways of preserving and keeping them for longer periods. In the process, humans exposed themselves to more potential pathogens, and disgust was one of the means of identifying and protecting oneself. As settled culture developed, disgust expanded beyond an instinct to protect the individual body into an emotion that could shape, identify, and protect the wider culture from the potentially malign influence of outsiders or the other—the idea that we as a group share values, including whether we do or do not like a particular food. Finally, disgust acquired a spiritual dimension, the notion that we simply do not consume particular foods. Defining conceptual, moral, and symbolic boundaries of behavior, this dimension helped humans to feel distanced from other animals as well as groups of people with different ideas.[57]

CHRISTIANITY AND SIN

Although the books of Leviticus and Deuteronomy are part of the Christian Bible's Old Testament, their food strictures are not generally adhered to by Christians. But this does not mean that food is insignificant to the belief system and practices of these groups. In the Christian model, the very act of eating, so closely tied to the original transgression—Adam and Eve consuming the forbidden fruit in the Garden of Eden—has, over time, been material for a discussion of sin that reached far beyond the deadly sin of gluttony. Although the biblical verses of Paul emphasize the importance of sharing food and of harmonious communal eating, thereby minimizing the relevance of food-based strictures, subsequent interpretation of Christian scripture gave more prominence to limits and rules.[58] In the fourth century, Jerome prescribed complete abstinence from meat, wine, and even most vegetables as the only protection against the heat of lustful desires, turning fasting into one of the key tenets of asceticism and, in a world all too familiar with the horrors of starvation, one of its most dramatic expressions.[59] In his thirteenth-century *Treatise on Fortitude and Temperance,* Thomas Aquinas points out the close link between gluttony and lust, excessive desire for food being a requirement of gluttony, which, at the same time, compounds the sin by being a species of self-harming lustful urge.[60] Despite the significant changes to the structures of faith that had taken place by the seventeenth century, aspects of these older interpretations of biblical writing continued to permeate the culture, and they persistently emerge in both religious and secular contexts into the twentieth and twenty-first centuries, frequently in relation to temperance and health foods or, as we'll discuss in chapter 11, in relation to the consuming body. Endowed with a moral value in a religious context, food is readily converted into an ethical symbol for alternative cultural practices and beliefs.

FASTING AND FEASTING

There are various religious practices that directly involve food, the most readily recognizable those in which humans exercise control over what is eaten through the observance of additional dietary rules or complete abstinence and those in which commemorative or ritual meals are eaten in celebration of a particular religious festival. Fasting and feasting are often linked, with feast following a period of selective fast, such as the celebration of Christian Easter after the forty days of Lenten fasting or the breaking of the total abstinence observed during daylight hours at sunset each day in the Muslim month of Ramadan.[61]

Whether or not a feast follows, fasts are usually undertaken to propitiate the gods and demonstrate atonement for sin or are practiced to mark a disastrous event in the history of the religion. In these cases, the fast is usually a day of commemoration and observance with no celebratory meal to follow, such as the Shi'ite fast of Ashura (which commemorates the martyrdom of Hussein in Karbala)[62] and the Jewish fasts commemorating the destruction of the temple and the exile into Egypt, Tish'a Ba'b, and Yom Kippur, the day of atonement. For Christians, every Friday was traditionally a day of Lenten eating, a kind of semifast in which no meat or dairy would be consumed and, in some denominations, no oil or wine. These same strictures would be followed every day during Lent. The most extreme ascetic practices are usually found within holy orders. Sufi mystics use *dhikir,* a prolonged process of sleep denial, recitations, prayers, and fasting to seek spiritual enlightenment, and similar practices are found in many religions. These practices might also be adopted by lay members of the faith as a demonstration of their intense religiosity. Caroline Walker Bynum discusses the adoption of extreme fasting by holy women in the Middle Ages as one of their key means of demonstrating their Christian religious credentials.[63]

Feast days celebrate key events in the history of religions, such as the Hindu festival of light, Diwali, which celebrates the achievements of Vishna and Lakshmi; Pesach or Passover, commemorating the Exodus of the Jews from Egypt; Mawlid, celebrating the birthday of Muhammad; Eid al-Adha, commemorating Abraham's sacrifice—averted at the last minute—of his son Ishmael;[64] and Christmas, celebrating the birth of Christ. Many of these, particularly those still closely bound to their religious roots, involve a ritual meal, such as the Passover Seder practiced in Jewish households.[65] Even those households not actively observing Christmas as a religious festival nonetheless tend to adhere to common cultural practices, including the composition of specific meals.

CONCLUSION

Mark Graubard tells a story, repeated by Frederick Simoons, of a young engaged couple—biologists—who went on a field trip together.[66] At some point in their expedition, the woman was startled and frightened when a large caterpillar fell onto her blouse. To reassure her that there was no danger, her fiancé laughed, plucked it from her—and then, to prove his point that it was "only protein"—ate it. The woman was shocked to the core by this action, as many readers of this volume might be. We don't know the precise nature of her horror, but we have seen that it could have stemmed from a number of sources.[67] Perhaps her faith prohibited the eating of insects or larvae; perhaps as a child—a budding biologist—she had collected caterpillars and watched them turn into butterflies and had considered them almost as pets; perhaps eating any still-living creature (rather than a properly processed one) was, for her, taboo; maybe she feared it was poisonous or had identified it as a rare species just as it disappeared down his throat; or possibly she'd simply never considered that anyone, let alone her future husband, might see a caterpillar as food, and her response was one of straightforward disgust. What we do know in this case is that the engagement was immediately broken off with the declaration that she "could not kiss a worm eater." With the help of some of the scholars explored in this chapter, we might be able to explain to the jilted consumer of larvae that although a single "rational" explanation might be difficult to find, he has a whole host of cultural possibilities to choose from. He might also have been reassured to realize that there's a very good chance that somewhere in the world he'd one day find a fellow soul who found his eccentric food habits completely culturally acceptable.

CHAPTER 5

Industrialization

Technology, Rationality, and Urbanization

How did we get to Spam? For some people, Spam—and the kind of mass-produced, canned industrial food like it—is a miracle. It helped sustain America, Britain, and Russia through World War II and has brought cheap, nutritious, and easily prepared food to millions of the poorest people in the world.[1] For many others, Spam is the epitome of all that is wrong with the modern food system. The industrial production and processing of food on a mass scale has been associated with environmental degradation and threats to public health, such as the Bovine Spongiform Encephalopathy (BSE) epidemic in the United Kingdom. Food movements like Slow Food argue that industrial food has eroded biodiversity and undermined gastronomic traditions and artisanal forms of production. The pleasures of making and eating food close to nature are being lost, replaced by bland uniformity and the "convenience" of factory-made products.

Such concerns about industrial food are central to many contemporary debates about food security, food movements, and public health, to which we will return elsewhere in the book. But we cannot understand these contemporary problems in full without placing them in a long history of food industrialization, a history that is the focus of this chapter. What we think of as "industrial food" in the present is the product of a complex set of interrelated technological, economic, social, and political transformations that have unfolded over the course of several hundred years. A familiar but flawed view of these processes sees

FIGURE 3. *Spam*, by Mike Mozart. *"Spam" by Mike Mozart is licensed under CC BY 2.0.*

the Industrial Revolution as a technologically driven event that gave rise to modern societies, fundamentally transforming social institutions and relations, including those involved in the food system. From this perspective, technological development industrializes agriculture, reconstructing societies in which economic activity has been centered on the production of food at subsistence levels to ones where work is oriented toward the manufacture of a diverse range of commodities of which food forms only a part. The food system, on this view, moves ever further away from nature and natural ways of living and becomes fully embedded in the social and economic processes that have denaturalized the modern world. More specifically, industrialization plays a central part in the commodification of food in the transition to capitalist society. Whereas prior to the Industrial Revolution, food had not been seen as a means of wealth accumulation, under industrial capitalism, the food system comes to be considered a means of profit.

This picture of the relationship between industry and agriculture, as well as the assumptions about the transition from the natural to the social that inform it, is too simplistic. There is much evidence that in the multiple factors that gave rise to industrialization on a large scale in England and elsewhere in Europe from the mid- to late eighteenth century, developments in the food system played a key role. Indeed, there are good grounds for thinking that fundamental changes to agriculture in the eighteenth century were an important determinant of the industrialization that followed. But in turn, these new developments in agriculture are closely related to the gradual rise of a preindustrial capitalist economy in western Europe, and in this regard, the rearrangement of cultural, social, and political relations was central.

Considering industrialization from the vantage point of changes to the food system allows us to gain a different perspective on the main question that has shaped social theory since the nineteenth century: the emergence of the modern world and the defining characteristics of modernity. While famous thinkers have addressed this question exhaustively, the consideration of patterns of food production, distribution, and consumption in the rise of the modern world has often been regarded a marginal or specialist exercise, something to be left to subfields in economic, social, or cultural history. As this chapter—and, indeed, the book as a whole—seeks to establish, the history, sociology, and political economy of food enhances our understanding of the modern world in important ways. The way in which we eat, and particularly the way in which food is distributed and exchanged, has transformed the physical environments of the modern world, shaping not just the public—and, indeed, "private" spaces of the city—but the rural landscape and the suburban transition from town to country. At the same time, as we will see in the next chapter, the physical spaces of food and drink consumption in towns and cities have played a considerable part in the formation of modern political ideologies, public and political institutions, and social movements. A proper understanding of "modernity" and processes of modernization cannot therefore overlook the significance of the food system in shaping the character of social, economic, and political relations in the present. If industrialization is a modernizing process (and there is room to question whether it is so per se), then it is as much a product of the transformation of the food system as the principal cause of that transformation. Accordingly, industrial food is not simply a consequence of economic industrialization driven by changing technology, but rather it is an outcome generated in the political, social, and economic relationships between people within and across societies.

In this chapter we pick up some of the core themes of the book—revolutionary change, the relationship between nature and society, the public and the private—by first reviewing some key accounts of the transition from agricultural to modern industrial society. We will see how the problem of the food system and its transformation is addressed in these accounts but also how it is seen as an effect, rather than a significant cause, of other processes. We will then turn to the Industrial Revolution, which begins in England in the 1760s and unfolds across western Europe and North America in the nineteenth century, analyzing its relationship to the food system in three periods. Finally, in light of the history of food industrialization, we will turn, in the conclusion, to industrial food in the present and, in particular, the impact on the food system on emerging forms of food biotechnology.

MODERNITY AND THE TRANSFORMATION OF THE FOOD SYSTEM

The three founders of sociology, Karl Marx, Émile Durkheim, and Max Weber, were all centrally exercised by the question of the rise of modern industrial society and the transition from an economic system ordered around agricultural production to one where economic activity is organized by industry. Yet it would be a mistake to think that concerns with the transformation of agriculture by industrial processes began with these figures. As Keith Wrightson claims, reflections on the growth of industry and the changing nature of agriculture stretch back to the sixteenth century in England.[2] Thomas More's *Utopia,* for example, contains a diatribe against the enclosure of arable land to support sheep for wool production.[3] But prior to the onset of the Industrial Revolution, the most extended reflection on the transition away from agriculture is provided by the theorists of commercial society who came out of the Scottish Enlightenment, such as Sir James Steuart, John Millar, and, most notably, Adam Smith. In his famous *The Wealth of Nations* (1776), Smith recognized the role played by food surpluses in generating demand for other goods: "Food not only constitutes the principal part of the riches of the world, but it is the abundance of food which gives the principal part of their value to many other sorts of riches." But food surpluses that thereby generate wealth are the outcome of "the improvement of the powers of labor in producing food by means of the improvement and cultivation of land."[4] Smith therefore places food production within a system for the generation of value through rent on land. Food, in that sense, is the source of rent, and the more cultivated and improved land is by labor, the higher the rent it can capture. The development of agriculture and the production of significant food surpluses is therefore necessary for the generation of other forms of wealth that come with the development of civilization, as was already signaled in chapter 2. Towns and cities, centers of manufacturing and trade, can only come into existence once agricultural activity has reached a level of productivity that allows for the sustenance of urban producers—artisans and craftsmen—who drive the development of manufacturing and the merchants who promote domestic and then foreign trade.

While Smith thus acknowledged the importance of the food system in the emergence of a commercial society defined by manufacture and trade, there are two aspects of his account that can be brought into question. First, Smith sees the modern manufacturing system as a development made possible by agricultural surplus, but which, once developed, radically changes the character of agricultural production. In this respect, while the town and country are mutually dependent, towns and cities come to have priority over the country and shape it to a much greater extent than vice versa. As we will see in the following section, it is not clear that in the early modern period in Britain, manufacturing in towns did play such a leading role. Rather, the divide between town and country was blurred, and the food system and the manufacturing system were deeply intertwined until the explosion of the urban population that came only in the wake of the Industrial Revolution in the nineteenth century. Second, Smith sees the move from agrarian to commercial society as given in the dynamic relationship between the increasing productive power of labor that comes with the division of labor and the growing population this supports. But, as Marx argues, this overlooks the role played by contentious social and political relations in the emergence of capitalism. Central to class conflict, in this regard, was the question of the control of agricultural land and the food system.

For Marx, capitalism revolutionizes the world by transforming social relations, making most people in capitalist societies dependent on money and market exchange to obtain the goods necessary to satisfy their needs and wants. Marx argued that a number of factors are involved in the development of capitalism in Europe: the general growth in trade that accompanied improvements in transport and communications from the fifteenth century; the emergence of new kinds of trade (including the transatlantic slave trade after the "discovery" of the Americas); and, as a result of the Spanish conquest of the Americas, an influx of new goods (see chapter 3) and, particularly, precious metals. This early modern period, which encompasses both what historians have called the "scientific revolution" and the "military revolution," as well as, later in the period, a new agricultural revolution, sees a rapid development of new forms of technology and the beginning of a profound transformation of rural and urban life. But Marx argued that capitalism did not have its origins in a benign expansion of trade and technology. Marx's understanding of "primitive accumulation," outlined in chapter 26 of *Capital,* involves a rejection of the notion put about by apologists for capitalism that it emerged from the thrift and industriousness of the few. Rather, "in actual history it is a notorious fact that conquest, enslavement, robbery, murder, in short, force, play the greatest part."[5] Capitalism depends not just on the rise of the bourgeoisie but also on a class of "free" workers who sell their labor power to the capitalist. These workers are free in two respects: they are not part of the means of production (they cannot be owned and directed like serfs or slaves), and they themselves do not own or control the means of production. This class has to be separated from the means of production—prior to capitalism, in feudalism, the workers were both means of production and had some degree of control over the means of production (land). "So-called primitive accumulation, therefore, is nothing else than the historical process of divorcing the producer from the means of production."[6]

The appropriation of land and the products of social labor as profit (what Marx calls "surplus value") is thus key to the rise of capitalism. It is also for Marx the precondition of industrial capitalism, which cannot take off without the prior commodification and monetization of agricultural products. Capitalism, then, begins in the food system, or at least it begins by extracting capital out of land used for agricultural production.[7] But the system for the extraction of surplus from food production would not have been possible without the development of a number of forms of knowledge and technical expertise that emerge in the early modern period. For Max Weber, capitalism was a "rational" system depending on particular kinds of innovations in calculation and accounting, such as the invention of double-entry bookkeeping, for the systematic accumulation of capital.[8] With respect to agricultural production, procedures for the calculation of crop yields and the profit to be made from them for future expansion went hand-in-hand with the development of the technologies associated with the new agricultural revolution, to which we will turn in the next section.

Weber saw such strategies of calculation in the development of "rational" capitalism as a reflection of a broader historical pattern in the West of "rationalization." The "disenchantment of the world"—the gradual erosion of religion and tradition as a ground for action—gives way to various modes of calculative action. For Weber, the central motor of rationalization in this sense is the modern state. Like Marx, Weber thus saw a key role for the state in the rise of capitalism, but unlike Marx, he considered the modern state to have emerged prior to capitalism not as an instrument to facilitate the accumulation of capital, but to secure a monopoly of political control of territory and populations. The rise of the sovereign territorial state in western Europe marks the period of transition from feudal social relations, which had largely disappeared by the fifteenth century, to the emergence of capitalism from the sixteenth century.[9] The exercise of a political monopoly over a given territory by the state creates the space for the development of a legal and administrative system that records and gives force to claims over land and productive property. More generally, as was the case in France under Louis XIV, the rise of absolute monarchs in control of the state tended to weaken the political and social position of the landed aristocracy, whose control of the land in the feudal system was a means of asserting military and political power against territorial rivals, rather than a means of wealth accumulation. The gradual decline of the aristocracy and the emergence of tenant farmers and a growing population of waged agricultural laborers contributed to the commercialization and monetization of the food system. This was further promoted by the security the state could guarantee over legal title and its increasing enforcement of contracts by its courts (as opposed to various local and regional courts presided over by the church and nobility). A common currency across a territory—which would eventually be guaranteed by state-controlled central banks[10]—as well as the standardization of weights and measures, provided stability in exchanges that further facilitated the commercialization of food production.

The accounts that Smith, Marx, and Weber give of the passage to the modern world all imply that the process involves a profound transformation in the character of the food system that occurs prior to the development of industrial capitalism. Their insights into this shift provide us with important conceptual and theoretical grounds for understanding the development of industrialized food in the contemporary world. For Smith, population growth and the diversification of the division of labor in towns and cities drives the improvement of agricultural land, boosting productivity and providing an impetus for technological innovation on the land. This orients the food system away from the land and toward towns and cities, where food markets, food processing, and public spaces of consumption increasingly develop. Marx's disclosure of "primitive accumulation" draws our attention to the way in which what today we might call "land grabs" are central to the creation of a private property regime based on commercial agriculture. The commodification and monetization of food occurs independently of industrialization, through the private enclosure of agricultural land and the appropriation of agricultural labor—both in Europe and its far-flung, particularly American, colonies—and was necessary for the birth of industrial capitalism in the late eighteenth century. Finally, Weber's emphasis on a broad process of rationalization in the rise of modernity demonstrates how the commercialization of food production, which paved the way to its industrialization, relied on the development of forms of calculation and the legal and political apparatuses of a modern state exercising a monopoly of authority over a given territory.

Despite the significance of their analyses, Smith, Marx, and Weber all overlooked the role of what we have called the "food system" as itself a driver of social, economic, and political transformations. Of course, historically increasing specialization, the commodification and monetization of land, and the development of forms of economic calculation and legal and political regulation of food production have strongly shaped the character of the food system. But we also want to explore the interrelationship between such processes and the food system itself. We can address this relationship further by examining the event central to the rise of industrialized food: the Industrial Revolution.

THE INDUSTRIAL REVOLUTION

Eric Hobsbawm has described the Industrial Revolution as the "take-off into self-sustained growth": "Sometime in the 1780s, and for the first time in human history, the shackles were taken off the productive power of human societies, which henceforth became capable of the constant, rapid and up to the present limitless multiplication of men, goods and services. . . . By any reckoning this was probably the most important event in world history, at any rate since the invention of agriculture and cities."[11] Hobsbawm also points out, writing in 1975, that the Industrial Revolution had not ended. Forty years later, that is still the case. Its locus may have shifted in recent decades away from Europe and North America toward East Asia and Latin America, but there is little question that at the global level, industrialization (and deindustrialization) is still going on. One key problem this perspective raises, however, is whether the changes of the past two hundred years can be seen as being part of *the* Industrial Revolution or whether it might be more revealing to think of a series of industrial (and technological) revolutions that have fundamentally altered social relations. The latter formulation is particularly useful for thinking about the relationship between industrialization and the food system. As we will see below, in the course of the changes that take place in production, transportation, communications, and other forms of technology from the late eighteenth to the mid-twentieth century, the food system is transformed a number of times. We will consider these radical changes in three "waves": the first wave, which sees the rapid development of the factory system in towns and cities; the second wave, in which steam-powered transport and developments in communication contribute to a remolding of the fabric of the city and its relationship to the countryside; and the third wave, in which mass production heralds the industrialization of food production and processing.

The Early Industrial Revolution

The first wave of the Industrial Revolution begins in the second half of the eighteenth century in the north of England and is usually seen as being sparked by two interrelated developments. The first is the invention of a number of machine technologies for the manufacture of textiles, powered by water turbines. The second is the concentration of such machinery, their power source, and the workers operating the machinery, in a single large building—at first the water mill and later the factory powered by steam and then electricity. In conventional accounts of the Industrial Revolution, it is these technological innovations that fuel the unprecedented productive growth that occurs in the late eighteenth and early nineteenth century and lead to the subsequent industrialization of agriculture.

The picture is more complicated, however, because this first wave of industrialization is made possible only by a series of transformations in social and economic relations in the early part of the eighteenth century. It is these changes that are closely implicated with developments in the food system. First and foremost, this period in Britain saw an expansion of population related to increased agricultural productivity. Innovations in crop rotation, most notably the adoption of the "Norfolk system," saw the widespread abandonment of the practice of leaving land fallow every third year. The Norfolk system involves the planting and harvesting of wheat, turnips, barley, and clover over a four-year cycle.[12] In addition to the Norfolk system, potatoes began to be more widely grown and eaten, providing a readier and more easily processed source of calories than grains such as wheat and rye.[13] A surplus of root vegetables produced fodder for livestock, allowing for the year-round supply of meat. Meat production and consumption increased considerably at this time,[14] with the British gaining a particular reputation for the cooking and eating of beef. *Les rosbifs,* the French nickname for the English, may have been first uttered in ravenous envy rather than chauvinist scorn, a view reflected in William Hogarth's famous painting *The Gate of Calais* (1749).

Developments in land management combined with new technologies to significantly improve crop yields in this period. Instruments such as Jethro Tull's seed drill and, later in the early nineteenth century, the first reaping machines considerably reduced the amount of time and labor required for sowing and harvesting, and they allowed for more reliable cultivation of large tracts of land.[15] But inasmuch as the British agricultural revolution of the late seventeenth and early eighteenth centuries was a necessary precursor of the Industrial Revolution,[16] it was itself a product of the profound social and economic changes occurring in Britain in the early modern period—most importantly, the rise of a capitalist national economy.[17] The major impetus for the increased productivity of agriculture was the appropriation of common land in enclosure. But while individual acts of enclosure could take place swiftly, as M. W. Flinn claims, "the legal machinery of enclosure must be formalized, and society conditioned to the acceptance of the social and economic upheavals involved."[18] Thus a key condition of the first wave of the Industrial Revolution was the simultaneous commercialization and regulation of the food system.

The factory system that characterizes the first wave did not, then, emerge overnight, and its initial impact on the food system was limited. As we have already seen in considering Smith's view of the relationship between town and country, there was no straightforward or rapid transformation from a society based on agriculture to one based on industrial manufacturing in the towns and cities. Rather, in early modern Britain, the divide between town and country was blurred. As Wrightson demonstrates, even at the beginning of this period, in the early sixteenth century, there was a significant nonagricultural sector in the countryside made up of manufacturing and trades.[19] This manufacturing sector, Maxine Berg argues, is considerably more developed by the time the Industrial Revolution takes off in the 1760s and is centered in homes and workshops, involving women and children as much as it did men.[20] The location of manufacturing production remained largely rural: the "putting-out" system involved the employment of craftspeople in the domestic setting or in modest workshops attached to homes, and it developed at a time when the urban population remained relatively small. Overall population growth in the eighteenth century prior to the Industrial Revolution had been rapid—between 1700 and 1801, the population of England and Wales increased from around 5.5 million to 9 million, and by 1831 it had reached 14 million.[21] The growth of the urban population began to accelerate as the eighteenth century moved on—it increased from 1.2 million to 3 million over the course of the century[22]—but was broadly in line with the general trend in the country. As Pawson argues, "the degree of urbanization in the eighteenth century was not great" and much of it was found in already large cities.[23]

If in the early Industrial Revolution, industrial urbanization did not radically alter the system of food production, it significantly changed social relationships around food. Long hours of factory work, poor wages, and cramped living conditions transformed the experience of everyday food preparation and consumption. Despite improvements in agricultural productivity, the mid- to late eighteenth century saw increases in food prices that led to a series of serious food riots in both the countryside and the burgeoning industrial towns.[24] Food would not become more plentiful or diverse for industrial workers until incomes rose in the latter half of the nineteenth century.[25] Most of the household wage was spent on a diet consisting, in the main, of thin soup, bread, porridge, potatoes, and beer, with the occasional availability of fresh fruit and small amounts of cheese or meat such as bacon and mutton. Scarcity in the food supply created hazards to health and life in the city: the adulteration of bread, tea, and other products was widespread, and whether from ignorance or design, little regard was given to hygiene. It was common to see bakers kneading bread in oppressive conditions, half-naked, with sweat pouring off them into great troughs of dough.[26]

In the slum conditions of early industrialization, men, women, and children went out to work, with the burden of domestic labor falling on women. The confinement of women to the domestic sphere would not take place until later in the nineteenth century, but the shift to the idea that domestic work was the responsibility of the wife and waged employment that of the husband emerges in early industrialization.[27] Early industrialization is not just responsible for the birth of an urban, industrial working class; it also transforms relationships in the city and lends itself to the development of urban bourgeois tastes and lifestyles that had been developing in the seventeenth and eighteenth centuries.[28]

As we will see in the next chapter, industrialization accelerates the development of a public sphere of consumption, one in which the political and public power of the bourgeoisie is consolidated, and through which an advanced commercialization of food and drink takes place.

The Second Wave of the Industrial Revolution

If the first wave of the Industrial Revolution is characterized by the development of a factory system and accelerated urbanization that only marginally affects food production, the second wave sees the industrialization of the food system and the reconfiguration of urban space as a site for the preparation, distribution, and consumption of food. The transition from town to country is also radically changed. Technologically, what drives this second wave, above all, is the application of steam power to transport. The development of the locomotive engine and the steamboat were the two most important innovations in transport in the modern era. Steam-powered travel led to the transportation of people and goods on a massive scale, shrinking the world significantly.[29] At the same time, almost instant communication between countries and across oceans was made possible by the building of an international telegraph network.[30] One important effect of this shrinking of time and space was to facilitate migration on unprecedented levels. Much of this movement of people was to fuel further industrial and infrastructural development, as was the case with the emigration of large numbers of Chinese to the United States to work on, among other things, the particularly perilous crossing of the Sierra Nevada in the construction of the Transcontinental Railroad.[31] These waves of immigration had an important impact on culinary culture, transforming "native" cooking as well as creating cultural hybrids—such as American-Italian or British-Indian cooking—that have become central to various national food cultures.[32]

The principal effect of the revolution in transport on agriculture in the United States was to shift agricultural space closer to the main arterial routes. Large farms, particularly those cultivating grain or raising livestock, needed to be close to important rail hubs, centered in towns and cities, or to inland or sea ports. In *Nature's Metropolis*, William Cronon charts how this reorientation of agricultural space around major towns and cities fundamentally transformed the character of Chicago and the American Midwest. Chicago's location gave it a natural advantage. As a principal city of the Great Lakes, it developed as an inland port, taking in agricultural products from the East. The arrival of the railroad radically cut journey times from the East Coast. By 1852, it took less than two days to travel from New York to Chicago, down from the fortnight or more it would have taken previously.[33] The transportation of grain and livestock more and more took place by rail, but also along watercourses. Flatboats, on which large amounts of lumber or grain could be shipped, were relatively inexpensive for farmers to construct, or in the Midwest they were able to book passage on steamboats, which significantly reduced the duration of journeys.[34] Over time it became less viable for individual farmers to cover the costs of transportation to processing plants in the city, and merchants would buy up crops and stock from many producers, shipping them to the city, where they would be sold wholesale.[35]

The growing movement of labor, commodities, and capital facilitated by the rise of mechanized transportation radically transformed both the rural and urban landscapes in the second half of the nineteenth century. Farming changed as a result of farmers having ready access to a wider range of agricultural technologies and resources made available in the city. Processed timber was used for the building of large wooden structures—barns, tool workshops, and so forth—that were seen as contributing to more efficient and expanded production.[36] The size of the farm increased, allowing it to meet increasing demand from growing urban populations for food. But the effects of agricultural "rationalization" were a de-skilling of farming, as it became increasingly mechanized and subject to scientific knowledge,[37] and the further migration of rural populations to towns and cities.

The growing diversification and commodification of agricultural produce in the second wave transformed the role of the city in relation to food, as it became a major site of food processing on an industrial scale. Chicago's Union Stock Yard became the model for a centralized meat trade, capable of rapidly gathering, slaughtering, and processing an astonishing number of animals. In 1868, it covered one hundred acres and could at any one time handle seventy-five thousand pigs and twenty-one thousand heads of cattle.[38] Food processing on this scale had very visible effects on the urban landscape and social relations. Upton Sinclair's *The Jungle,* written in 1906, is a classic social novel that exposes the treacherous working conditions of the stockyards, which employed predominantly migrant workers.[39]

Inland and sea ports processed grain continuously, with ships offloading their cargo into large mechanized grain elevators, where it would be graded and packed before being stored or loaded back on to ships for export.[40] This expansion of waterborne trade, much of it in food, would be a major driver of American urbanization in the late nineteenth century, rather than industrialization per se. Elsewhere, as well, international trade triggered the emergence of large cities as trade hubs. As Jürgen Osterhammel claims, port cities tend to be treated as peripheral and disruptive,[41] but for exactly that reason, they played a key role in transforming social relationships around food as well as in accelerating the import and export of different kinds of food, drink, and cooking styles. The international network of port cities involved in global exchange expanded beyond the industrializing West and, through colonial expansion, reached as far as Mumbai, Shanghai, and Hong Kong. Such cities became sites of cultural hybridity, developing distinctive food styles that would be exported back to the colonizing power.

The shrinking of time and space engendered by new transport and communication technologies led to unprecedented levels of food exchange across the world and increasingly rendered societies dependent on an ever more complex international food system—the "food regimes" to be explored in chapter 10, on political economy—which made national self-reliance effectively untenable. This afforded sections of the population in industrialized or industrializing societies, mainly relatively wealthy urbanites, access to a hitherto unknown variety and abundance of food and food styles, but it also left many more dependent on a narrow range of food commodities, the production and quality of which was increasingly beyond their control.

The Third Wave of the Industrial Revolution

The true age of industrial food—the third wave of the Industrial Revolution—only arrived in the twentieth century, with the application of new mass manufacturing techniques to food production. It was another innovation in transportation, the automobile, that provided the paradigm case of this form of manufacturing. Henry Ford did not invent the automobile or the form of assembly-line production that today bears his name, but both the car and the Fordist system of production transformed social and economic relations in the early twentieth century, impacting the food system in the spheres of both production and consumption.[42] Mass production on the assembly line was the key condition for the development of "convenience" food, highly processed products that could be relatively easily prepared and served in the domestic kitchens of twentieth-century industrialized societies. Yet it was not this form of production alone that accounted for the ubiquity of industrial food in the twentieth century, for it presupposed other developments in food technology on the one hand—in particular, advances in preservation and substitution—and the acceleration of food commercialization on the other, which came with the emerge of the food corporation and mass advertising.

The two main innovations in preservation technology exploited by mass production in the first part of the twentieth century were canning and refrigeration. The principle of preservation by means of sterilization was first realized on a large scale by the French confectioner Nicolas Appert. Appert discovered it was possible to preserve perishable items by sealing them hermetically in glass jars and immersing them in boiling water for a certain length of time.[43] Appert's explanation for the process was that the heat of the water destroys the "fermentations" that led to spoilage by driving the air out of the jars. Only later would Louis Pasteur demonstrate that the anaerobic organisms that caused putrefaction could live in an air-free environment, and that this form of preservation, later called "pasteurization," worked by heating the food to a sufficient temperature and for a sufficient length of time to kill off most of the microorganisms that cause spoiling.[44] While Appert began production of food preserved in glass on an industrial scale in 1804, it was Bryan Donkin's replacement of glass by iron tins in England in 1812 that allowed for the possibility of cheaply produced preserved food en masse. Yet for much of the nineteenth century, mass production of canned food was limited, and it was used primarily on long sea journeys.[45]

While "icehouses" using natural ice had been the preserve of the wealthy at least from Mesopotamian civilization onward,[46] it was only with the development of mechanical refrigeration in the nineteenth century that preservation by freezing and chilling became widespread. The cooling effects of evaporation had been known since the mid-eighteenth century, and in the early nineteenth century, Oliver Evans described the action of the vapor-compression cycle, which allowed for the constant evaporation and, by means of compression, liquefaction of a gas in a sealed system. This was, in effect, a heat pump, expelling heat to the outside of the device while maintaining the circulation of a cooling refrigerant gas on the inside. Modern fridges and freezers of all sizes continue to employ this technology.[47] But before the development of the domestic fridge and freezer, it was used primarily in shipping and on the railways to expand supply chains, particularly in the shipment of meat, which could now be kept much longer without recourse to preservation in salt. The domestic refrigerator today is only one part of a much more expansive "cold chain" that makes possible a global trade in chilled and frozen goods on an unprecedented scale.[48]

Canning and refrigeration allowed for the prolongation of the life of perishable foods, mainly meat, fish, and fresh dairy products. But food substitution offered another way to deal with the problem of perishable or expensive food. Two prominent nineteenth-century discoveries that would later be produced on a mass scale were refined sugar from sugar beets and the invention of margarine. To refine sugar from sugarcane is a relatively easy process, but the disadvantage of sugarcane is that it flourishes only in warm temperate and tropical conditions. Sugar beet, in contrast, can be grown in colder environments, though it requires much more processing. Faced with a blockade of its ports by the British in the Napoleonic wars, Napoleon sponsored the cultivation of sugar beet in Silesia and oversaw the introduction of a number of factory refineries in France.[49] Later on, in the Second Empire, the French navy's demand for a butter substitute led to the synthesizing of margarine, originally an emulsion of animal fats and milk and later of vegetable oils and skimmed milk or water.[50]

As we have shown in this chapter, transformations of the food system that occur with industrialization cannot be explained away by techno-logical innovation alone. It is important to consider the political, eco-nomic, social, and cultural drivers of technological change, as it is the role of the food system itself in promoting or impeding the trajectory of industrialization and urbanization. The advent of mass-produced indus-trial food in the twentieth century needs to be seen in this same way. Technologies of preservation and substitution are employed alongside the "rationalization" of the work process, as seen in the assembly line and approaches based on the scientific management of labor time and movement.[51] The imperial expansion of Western states through military conquest and trade rested on technologies of mass food production and distribution. The extension of supply lines for armies and navies, the promotion of peripheral dependence on the metropole, and the export of domestic industries to colonies were all made possible by the growing industrialization of the food system. Within states, it is possible to see how mass-produced industrial food played a role in cementing nation-alist cultural identities and allowed states to more effectively regulate and control the health and welfare of industrialized and largely urban-ized populations (see chapters 7 and 8).[52] An industrialized food system both facilitates and is further advanced by the emergence of the interna-tional food corporation. Corporations like Tate & Lyle and Coca-Cola were formed in the food regimes that linked together mass industrial capitalism, colonialism, and global trade.[53] Companies like Kellogg's were instrumental in developing a marketing and advertising industry that by the mid-twentieth century formed an integral part of the food system in contemporary capitalist societies.[54] If we return to the question posed at the beginning of the chapter—how do we get to Spam?— we can see clearly that this is a process as complex as those responsible for the shaping of the modern world. Across the world, industrialization is far from a finished process and continues to fundamentally reshape food systems and cultures. But in a world where the technologies of industrialization are constantly transformed in their relationship to the ever-shifting political, economic, social, and cultural relations of contemporary capitalism, what is the future of industrial food?

CONCLUSION: INDUSTRIAL FOOD BEYOND SPAM?

On August 5, 2013, Professor Mark Post of Maastricht University in the Netherlands presented the first in vitro hamburger to the world. The meat for the burger had been developed in a laboratory using myosatellite cells, a form of stem cell, cultivated in a petri dish with a nourishing medium treated with bovine fetal serum.[55] The meat was grown over three months at a cost of US$325,000, funded by Sergey Brin, a cofounder of Google.[56] While clearly commercially unviable at this level of cost, the backers of in vitro, or "cultured," meat claim that with the right amount of investment and further technological development, it can be made at an equivalent if not lower cost than conventionally produced meat from livestock. Given that the process is self-replicating— once the meat has developed in the laboratory, cells can be harvested from it to grow more meat—it will not be reliant on livestock for further growth. The promise of this technology for people opposed to mass industrial farming—which they argue involves environmental destruction and cruelty to animals on a massive scale—is clear.[57]

At the moment cultured meat can be grown only in strands, but in this form it could clearly act as a substitute (cost and taste considerations notwithstanding) for highly processed meat products that involve fine mincing or pureeing of meat. Spam is one such product. Should we, therefore, be running with open arms toward cultured meat and other modern forms of food biotechnology as a means of improving animal welfare, protecting the environment, or even improving public health? Perhaps not. But the reason we should be skeptical about the claims made for in vitro meat and for technologies such as genetically modified crops is not because they invite the specter of Frankenfood—technologies that run away from our control and produce a nightmarish world of environmental collapse and epidemic disease—but rather because they tend to be made in highly simplistic terms that overlook the embeddedness of such technologies in a complex array of social, economic, and political relations.[58]

Cultured meat and genetically modified crops draw on technologies that stem from the "molecular revolution" in biology that followed the discovery of the structure of the DNA molecule in 1953. That revolution has fueled the emergence of forms of biotechnology that today— alongside other developments, such as the exponential growth of computing power, the increasing sophistication of robotics, and the rapid improvement of machine intelligence—promise to introduce a fourth wave of the Industrial Revolution.[59] The impact of these changes on the conventional food system is already clear, as genetic modification combines with technologies such as hydroponics to facilitate the growth of disease-free crops in closely controlled environmental conditions that allow for out-of-season production. But what the history of food industrialization explored in this chapter demonstrates is that industrial food technologies are not neutral; they are formed and applied to meet definite political, economic, and social objectives. We are right to be skeptical of claims that a new industrial revolution in the food system will create a better world, because the old industrial revolutions in the food system have, by and large, been used to consolidate the power of states over individuals, to promote exploitative colonization, and to pursue profit at the cost of human happiness. That is not to say that the food system in the modern world has not brought with it clear benefits, such as reducing the burden of agricultural labor, improving nutrition for many of the world's poor, and providing an unprecedented variety of foods and cuisines for very large numbers of people to enjoy. But whether the food system of the future will benefit the many or the few is a matter of politics rather than of technological innovation.

CHAPTER 6

The Public Sphere

Eating and Drinking in Public

An early entry in Samuel Pepys's celebrated diary tells us that on January 10, 1660, he spent the morning in a London alehouse with the mathematical instrument maker Ralph Greatorex. After conducting business, at four o'clock he went with acquaintances to "a cook's dinner" and then, having returned from some more business, went with a Mr. Jennings to drink "a pint of wine at the Star in Cheapside." From there, he traveled to Westminster, "to the Coffee-house where were a great confluence of gentlemen; viz. Mr. Harrington, Poultny cheareman, Gold, Dr. Petty, & c., where admirable discourse till 9 at night." Afterward, he went with his colleague Thomas Dolling to the Lamb's tavern, before returning home briefly to write a letter and then ending the day in Harper's tavern.[1]

What is striking about this not untypical entry is not the disclosure of the sheer volume of food and drink that Pepys and his contemporaries consumed, but rather its revelation of the way in which they conducted their affairs (in more than one sense of the word) in public. Much of Pepys's life was played out in the presence of others, and his "private" self was defined in terms of its public manifestation; as a modern phenomenon, the genre of journal writing depends in the first instance on such a private revelation of the public self, rather than the public revelation of a self constituted in private.[2] Today, while we tend to make clear distinctions between relationships fostered in the formality of the workplace, friendships played out in shared spaces of sociability,

FIGURE 4. *Edward Lloyd's Coffee House*, by William Holland, 1789. *Wikimedia Commons, https://commons.wikimedia.org/wiki/Lloyds-coffee-house-london-by-william-holland.jpg and permission tag is PD-Art, PD-old-100*

and the intimate relationships of families and lovers in the private domain, fluidity across these kinds of relationship was considerably more marked in Pepys's milieu. The rise in seventeenth-century Europe of the "public sphere" sees not so much the invention of novel spaces of consumption—the coffee shop was relatively new to London, but the alehouses and taverns had existed for centuries—but rather their transformation into sites in which these emergent forms of social and political relationships developed. Pepys's public sphere was, with little doubt, bourgeois in character, and the physical spaces of bourgeois engagement were inextricably tied to transformations taking place in the food system in the course of the development of capitalism, the modern state, and the accompanying industrial revolutions. But it is possible also to discern the emergence of a plebeian public sphere—and perhaps later, with the advent of industrial capitalism, a proletarian public sphere—whose fortunes were also intimately connected to the changes taking place in patterns of food production, distribution, and consumption that we have already charted. As we will see in this chapter, the centrality of spaces of consumption for the rise of the public sphere—or perhaps more accurately, for the rise of public spheres—cannot be decoupled from concurrent transformations to the food system explored in previous chapters.

In this chapter, then, the relationship between the public sphere, conceived of as both an idea and a distinct set of material institutions and practices, and the changing food system in modernity are considered. Three different spaces of congregation and consumption are examined: the alehouse, the coffeehouse, and the restaurant. The aim of this investigation is to see how we might better understand the key distinction in social and political theory between the public and the private through a consideration of shifting practices of eating and drinking in "public" spaces and, particularly, how this relates to actual and potential forms of political action.

THE PUBLIC AND THE PRIVATE

The notion of a distinction between public and private activities and public and private modes of living can be traced back to classical Greece and Rome. For the Greeks, the *oikos* was the sphere of the private—the home, in which all material production related to the satisfaction of biological needs took place. In contrast, the public was the *polis*, the city as a site of active engagement between citizens. Its physical location was the marketplace, the *agora*, where disputes and contests were played out, as well as where men resolved in concert to do great deeds—usually in the guise of warfare—for the glory of the city. But if the homestead was indeed the site of food production, which took place largely outside the material spaces of public interaction, the consumption of food and drink straddled the realms of the *oikos* and the *polis*. The *agora* was the setting for civic government and public worship, a significant aspect of which was the celebration of religious feasts. The common meal thus had both civic and sacred properties, Aristotle among others outlining its democratic character and noting how it was made accessible to the poor in Crete by public funding.[3] At the same time, the *agora* was a space of commercial exchange and everyday consumption. Merchants would operate out of the *stoas* that bordered the square, selling meat, fish, vegetables, and wine.[4] In the Roman Republic and Empire, while specialist food markets existed, the purchase and consumption of food and drink took place in numerous spaces that crossed the line between the public and private domains, most notably the taverns, which were often no more than private houses with dedicated spaces made open to customers.[5]

The close proximity of sites of public consumption to the stages of economic, political, and religious performance did not disappear with the waning of the classical world, and the European city in the Middle Ages, particularly in Italy, continued to structure civic space around the antique distinction between the public and the private. Yet with the fall of the Roman Empire, the city fell into decline as the principal source of political authority, with the feudal system fusing private ownership of land with political power. The domination of the landed nobility and the development of a concentration of power in the hands of monarchs who controlled large territories as fiefdoms confined lawmaking and politics to the courtly life of the nobility and monarchy. In chapter 8 below, we show how Norbert Elias's work on the "civilizing process" establishes that this replacement of the public civic sphere by the court was a prelude to the emergence of a form of mannered "civility," including the refinement of cooking and the promotion of etiquette at table, in a courtly society that would provide the resources for the development of a form of bourgeois civility central to the emergence of what Jürgen Habermas calls "the bourgeois public sphere."[6]

Habermas's famous book *The Structural Transformation of the Public Sphere* charts the rise of this arena from the time of the emergence of financial and merchant capitalism in the northern Italian city-states in the thirteenth century. The traffic of commodities and news set in train by early capitalism would later come to be organized through the depersonalized authority of the modern European state, which subjected commercial activity in "civil society" to its political and legal order. Importantly for Habermas, however, the rise of state-regulated capitalism and a bourgeois commercial or civil society is not simply a matter of the perpetual growth of commodity exchange but also of an incessant development in the exchange of information through the vehicle of the press. The private exchange of news between merchants and financiers—concerned with the general conditions affecting trade—came to take on a public character as that news was reported in print. This meant that "the traffic in news developed not only with the needs of commerce; the news itself became a commodity."[7] From the outset, then, the news had a commercial character, but it also came to have a political aspect insofar as the state sought to regulate and control the news media (and particularly any quarter of dissent) and to make itself public through the statement of official positions and policies in newspapers and journals.

The bourgeois public sphere was thus a literate sphere in which not just information but also opinions between private people were publicly exchanged. Crucial for this exchange was the development of spaces of congregation and discourse, particularly in urban centers. Habermas points in particular to coffeehouses, the salons, and the *Tischgesellschaften* (table societies in Germany) as representative of these physical spaces.[8] In the London coffeehouses, the aristocrats who mixed with the bourgeois intellectuals still (unlike in France) "represented landed and moneyed interests." Accordingly, "critical debate ignited by works of literature and art was soon extended to include political and economic disputes, without any guarantee (such as was given in the salons) that such discussions would be inconsequential, at least in the immediate context."[9] As a site of material consumption, the coffeehouse came to take on a central political function in the bourgeoisie's colonization of state power. Habermas also thinks, however, that the very form of the discursive exchanges that went on in the coffeehouse and the other sites of the bourgeois public sphere—the conversation—provides the grounds for the exercise of a much more extensive form of emancipation in modern societies. But as we will see presently, such a claim overlooks the significance of material spaces of consumption for the articulation of a transformative politics in the seventeenth and eighteenth centuries, while at the same time glossing over the decidedly non-emancipatory forms of relationship that were formed in and as a result of relationships developed in the bourgeois public sphere.

The political theorist Hannah Arendt, in a passage that would seem consistent with the denial of the specifically political character of eating and drinking together, claims that the "sociability arising out of those activities which spring from the human body's metabolism with nature rest not on equality but on sameness."[10] Arendt is making the point that equality, in terms of the articulation of differences, comes only in the context of political association. A political association expresses not identity but plurality, whereas round the table where we eat and drink with others in public, we are all the same—consumers of the items placed before us. This view seems consonant with that of Habermas to the extent that he would consider acts of eating and drinking as merely contingent to the development of the political public sphere. Yet while Arendt is not greatly concerned with eating and drinking as features of the human condition, happy to confine them to the category of the biological requirements of the *animal laborans,* her emphasis on participation in public life as a form of performance draws us away from this austere interpretation of public consumption. Public eating and drinking can be seen as acts of performance that are inextricably tied to the other forms of display involved in the ascendency of the bourgeois public sphere. This performativity of public consumption informs Richard Sennett in his book *The Fall of Public Man,* a work that shows both continuities and departures from Habermas's view of the bourgeois public sphere.

Sennett argues that the eighteenth-century coffeehouse did indeed involve a form of equality (though one between men, with women being largely excluded) that has, in Arendtian terms, the character of politics or free action. The enjoyment had by the patrons of the coffeehouse was found in conversation with other people. The central rule governing speech was that "distinctions of rank were temporarily suspended; anyone sitting in the coffee-house had a right to talk to anyone else, to enter into any conversation, whether he knew the other people or not, whether he was bidden to speak or not." Coffeehouse speech was thus a "sign system of meaning divorced from ... symbols of meaning like rank, origins, and taste, all visibly at hand." At the same time, the conventions of conversation meant that people engaged in sociable activity were not compelled to reveal much about themselves as persons. Strangers could "interact without having to probe into personal circumstances."[11] Sharing food and drink at the table were essential aspects of establishing relationships of sociability independent of rank: to consume the same food and drink was not to be the same, but rather a necessary condition of the expression of difference through opinion. The material spaces and practices of consumption in the coffeehouse (and in the alehouses and taverns) were not, therefore, incidental to the performance of the free activities of argument, but a condition of them.

Sennett shows how public acts of consumption could therefore be revelatory of the qualities of "private" persons, but where privacy is a quality understood in a quite different sense to the privacy of the modern self, a character that comes to the fore from the late eighteenth century, and which, for Arendt, marks the transition in the industrial capitalist world to a society of consuming selves that come to dominate public life. In Arendt's terms, this represents the reduction of politics and the public realm to "society"—understood as the collection of selves, united in their sameness as creatures of pure consumption.[12] In modern societies, eating and drinking become less exercises in public performance and more vehicles for the expression of the identity of consuming selves. This occurs in a wide variety of ways, such as the reestablishment and representation of rank according to the "refined" tastes of elites; the disdain of food and drink as dangerous to corporeal purity, and thus being subject to strict self-regulation; or, in more recent times, the aestheticization of certain kinds of food and drink preparation and consumption among the fashionable young avant-garde. Whatever these forms of the expression of identity through food take, they have become removed from spaces of consumption in which public participation was itself seen as the primary goal and the sharing of food and drink its vehicle. In the remainder of this chapter, we monitor this process of the "privatization" of the public character of eating and drinking by considering three institutions whose publicity has been repeatedly transformed in the last four centuries: the alehouse, the coffeehouse, and the restaurant.

FROM THE ALEHOUSE TO THE PUB

The drinking-house is a global phenomenon that has appeared in many human societies over the course of millennia. As we saw in the previous section, taverns were common in classical antiquity, and some form of shared drinking spaces are likely to have existed in towns and settlements from the time of their foundation at the beginning of the Neolithic age, simultaneous with the earliest production of wine and beer. Premodern drinking-houses appear in many different configurations, from makeshift dens to large communal halls. In all of these guises, drinking-houses have appeared in some way as public spaces, at least in the minimal sense that they are not private. If not open to just anyone (outsiders, women, children, slaves, and so on, may be banned), members are not necessarily related by blood or by emotional bonds, and they may well be strangers. But for our purposes, the publicity of a "public sphere" has to go beyond the contingency of the relationships played out in the premodern drinking-house. As Sennett's argument suggest, the "public house" has to be understood as a space in which social rank and distinction is replaced by a certain kind of equality established within the performance of public activity. In that regard, a "pub" has to be more than a space in which people drink (and get drunk) together; rather, it requires a specific set of social relationships. We may think that such social relationships are a condition of modernity, but this is a questionable supposition, given that the substantive egalitarianism of certain forms of drinking-house predates the onset of the modern world and occurs in places outside the West. At some point, however, and in a range of registers encompassing activities that are simultaneously economic, political, and cultural, the drinking-house in the West comes to develop into a space in which a substantive publicity beyond sociability becomes established. A central feature of this transformation is that the drinking-house itself comes to be seen as a public space, a site in which engagement with others happens in distinction to private activity and which is necessary for the formation of civic and political identity.

After the decline of the tavern in classical antiquity, the drinking-house reappeared in significant numbers in Europe only in the High Middle Ages, that is, starting around the eleventh century. By the early fourteenth century, it had become the main provider of hospitality.[13] In England and elsewhere in Europe, there developed a three-way division between different types of drinking-house: the inn provided food and drink for travelers, as well as lodging and stabling for horses; the tavern served its customers wine and food and was frequented mostly by the upper and middle ranks; and the alehouse was the most basic form of drinking-house selling ale primarily to urban artisans, household servants, and agricultural workers.[14] As Mark Hailwood shows, the latter kind of public drinking establishment had its origins in the use of hops to flavor ale. Hopped ale—beer—had the advantage over spiced ale that it could be kept for periods of weeks rather than days before going off, since hops act as a preservative. This discovery acted as a spur to commercial brewing and took both the production and retail of ale out of the hands of small domestic brewers.[15] As we see in the following chapter, a second development in drinking in the late medieval period—the increasing consumption of spirits—also boosted the commercial development of drink and drinking establishments and contributed to the development of new public sites of imbibing.[16] By 1700, there were an estimated fifty-eight thousand alehouses in England alone.[17]

Commercial imperatives, then, clearly had a role in the emergence of the public house, though they do not exhaust the reasons for its rise. As Hailwood shows, levels of beer consumption may well have been falling in the early modern period, perhaps quite substantially.[18] There may, however, have been increasing demand for alehouses as recreational sites, given the suppression of alternatives. Communal feasts and festivals that had taken place in the churchyard (with the church's sale of beer providing a valuable source of revenue) were suppressed after the English Reformation by Protestant clerics and politicians, while the dissolution of the monasteries (1534–39) under Henry VIII had already diminished another source of ale for public consumption.[19]

The rise of the alehouse as a recreational space, rather than simply a site of commercial exchange, prefigures a key feature of public spaces of consumption in the modern period. For Hailwood, the "good fellowship" sought out and enjoyed by drinkers points to the significance of the alehouse as, first and foremost, an arena of sociability. In the seventeenth century in England, this sociability would become a vehicle for political action. In the course of the conflict that would lead to the English Civil Wars of 1640–49, the taverns—where wine was the main drink served—became the focus for Royalist drinking rituals that would involve drinking the health of the king with wine. This kind of "loyal-health" toasting had religious origins in the Eucharist,[20] but it came to have a political function, allowing for assembly in favor of and expression of loyalty to the king. In contrast, the king's parliamentary opponents—particularly the Puritans—expressed their severe disapproval of public drinking, and during the interregnum, there was a crackdown on drinking and drinking establishments. After the Restoration, those who refused to participate in public drinking, largely the abstemious Puritans, were regarded with political suspicion. To not drink implied sedition, and to congregate in any kind of drinking-house, inn, tavern, or alehouse was seen as a sign of the affirmation of the political and social order established with the return of the king.[21]

By the beginning of the eighteenth century, then—the point at which the term *public house* began to be widely used[22]—the elements of the publicity of drinking-houses had been put in place. We see establishments that are not simply private spaces opened up to all comers, as was the case with the medieval alehouse, but places where distinct forms of sociability manifested in drinking rituals reflect social and political identities and divisions. However, the development of the pub in England in the eighteenth century sees both further commercialization and the gradual dissolution of the illicit alehouse (and dram shop, as noted in chapter 7). Increasing licensing and regulation of drinking-houses by the state also began to transform their character.[23] The building of specialist public houses now included the differentiation of internal space to separate people by class and gender.[24] In terms of gender, this marks a striking contrast to the era of the "gin craze" in London, when women and men had drunk openly together (see chapter 7). To be sure, inns, taverns, and alehouses would carry on in their established form until well into the nineteenth century, but the industrial revolution saw the establishment of the pub as the predominant form of drinking-house. The "gin-palaces" of the nineteenth century were the model for large pubs in towns and cities, with the remnants of the one-room alehouse being relegated to small towns and back streets where, in the twentieth century, they would become overwhelmingly the domain of working-class men.[25] "Public bars" and "saloon bars" segregated working-class and middle-class drinkers, and many establishments even had dedicated "ladies bars" in which women were expected to drink. Thus, the pub as a form of architecture, typically seen as embedded deep in England's heritage, is, in fact, a relatively recent invention. It is perhaps an irony that the "pub," as such, came through the twentieth century to be an increasingly de-publicized space, as its functions returned to sociable (and in the late twentieth century, its moralizing critics would add, an increasingly unsociable space in which binge drinking, violence, and sexual debauchery were encouraged) drinking, and its civil and political functions were eroded. Certainly, the notion of the pub as a site of a plebeian public sphere in this later period is misplaced, even though such a sphere may have been said to be constituted in England in workingmen's clubs.[26]

CAPITALISM AND THE COFFEEHOUSE

As we have seen, one way of interpreting contemporary theories of the rise of the public sphere is that while distinct material spaces of congregation are one of its central features, what is consumed is only a secondary matter. What characterizes those spaces is the forms of action that take place in them, with a particular emphasis on the way in which the constitution of the class identity of the bourgeoisie comes about through the media of speech and argument. This construal of the social function of the drinking-house in general is not, however, confined to the commentaries of contemporary social theorists. Noting the prohibition on the consumption of alcohol in Islam (at least for the common people drinking in public), the first Western chroniclers of the Turkish coffeehouse in the seventeenth century considered coffee as a substitute for wine or beer.[27] The places in which coffee was consumed fulfilled much the same sociable function as the alehouse or tavern. As George Sandys, an English visitor to Constantinople wrote: "Although they are destitute of Taverns, yet have they Coffa-houses which something resemble them. There sit they chatting most of the day; and sippe a drinke called Coffa."[28]

Yet the history of coffee demonstrates its unique impact on social relations and institutions, and dispels the idea that it (or for that matter, specific forms in which wine and beer have been produced and consumed) is epiphenomenal for the development of the public domain in the modern world. As Brian Cowan has argued, those early English commentators on the Turkish coffeehouse were mistaken about its affinity with the kind of public spaces of sociability developing back home in the seventeenth century. The Turkish coffeehouse was, first and foremost, a site of local sociability, whereas the tavern and the coffeehouse, as they would develop in Europe, and particularly in London, had a cosmopolitan character.[29] That does not mean that distinctive national coffeehouse cultures did not develop, but the kind of cosmopolitanism of the London coffeehouse did become something of a model for the development of coffeehouses in the rest of Europe. What might be said to be particularly significant about the coffeehouse in London was the role played by the cultural *virtuosi,* independent thinkers (rather than courtiers) associated with institutions like the Royal Society, who placed much value on particular forms of manners and civility in conduct.[30] The peculiarity of the cosmopolitan model developed in London was twofold. First, in the figure of the *virtuosi,* it involved the social fusion of the aristocracy and bourgeoisie. Second, the London coffeehouse now came to play a central role in the emergence of a cosmopolitan capitalism that superseded the agrarian capitalism that had been developing in England since the fourteenth century. In a famous essay, Perry Anderson argued that this fusion of the aristocracy and bourgeoisie, encapsulated in the signal moment of the English Revolution during the Civil Wars of the 1640s, was a peculiarly English phenomenon, distinct from the revolutions in Europe from the late eighteenth century onward, in which the urban bourgeoisie vanquished the landed aristocracy. This fused English ruling class ruled not simply by virtue of controlling the coercive apparatus of the state but also by exercising a form of cultural hegemony. If this characterization of the developmental path of the ruling class is correct, it highlights the importance of the institutions through which its cultural superiority was expressed, the café being central among these in the seventeenth and eighteenth centuries. But it also sheds light on how such spaces of consumption could themselves become so central in the development of a form of cosmopolitan capitalism that represented a new stage in capitalism's trajectory, as its focus shifted from the commercialization of English agriculture to the building of networks of trade and finance across the globe.[31]

The importance of the English coffeehouse in the rise of cosmopolitan capitalism can be seen in the shift in its composition and in judgments of it by its own patrons between the seventeenth and eighteenth centuries, in the period between the eras when its two most famous *virtuosi* left their impression on its history: Samuel Pepys and Samuel Johnson. As Sennett remarks, this shift involves a reconceptualization of the coffeehouse from being a space of sociable performance between equals that was open to all "the public" to an exclusive and clubbable domain of the gentleman whose social graces and standing were brought in from the outside.[32] This privatization of the coffeehouse (recognized by Habermas in his account of the colonization of the "lifeworld" by the "system") sees the gradual diminution of the coffeehouse in public life in the late eighteenth century, a movement replicated elsewhere, such as in the famous Café Procope in Paris.[33] But the decline of the coffeehouse as a site of public engagement took place also from within, with the segregation of clientele by interest and occupation, a condition most evident in the London's burgeoning mid-eighteenth-century coffeehouses centered around the Royal Exchange.

The coffeehouses around the exchange became prominent from the late seventeenth century onward. Trades that would take place on the floor of the Exchange spilled out into the coffeehouses, which increasingly became sites for specialist financial news—in particular, the posting of prices in companies trading in joint stocks. Acts of intermediation between buyers and sellers—the job of the stockbroker —originated in the coffeehouse, and specialist deals in securities conducted by jobbers also began in their confines.[34] The lucrative market in shipping insurance started in Lloyd's Coffee House (see William Holland's 1789 illustration at the head of this chapter), which lent its name not to a company—Lloyd's of London—but to a market in insurance, created and maintained by an act of Parliament; and Jonathan's Coffee-House would become the site of the London Stock Exchange.[35] The coffeehouse thus increasingly lent itself to the accommodation of specialist trading and financial interests, catering to a cosmopolitan clientele. It expressed a complex system of social and political division that was nonetheless cut across by the common commercial interests of the parties involved.[36] As they developed as specialist sites of congregation, the exchange coffeehouses became increasingly exclusive, drawing up regulations banning day traders and eventually introducing membership. A similar trend would develop around the coffee shops of Wall Street in New York, with the Tontine Coffee House there acting as the foundational institution for the New York Stock Exchange.[37] Of course, one of the key commodities that the traders dealt in was coffee itself, and by the mid-eighteenth century, London was at the center of the global system in the production, distribution, and consumption of coffee. That system was woven into the colonial expansion of the West considered in chapter 3, as coffee found new main sites of cultivation in Central and South America, Southeast Asia, and across Africa.[38] The globalization of capitalism was in a key respect "fueled" by coffee consumption. The coffeehouses facilitated the rise of trade in coffee and other commodities (including tea, sugar, tobacco, and slaves). The shipping insurance and other insurance markets developed in London provided the security for the risks involved in the expansion of production and trade, and this growth fed further rises in the consumption of coffee across the world in the nineteenth and twentieth centuries, even if in England the greater commercial advantages gained by the trade in tea and sugar by the East India Company led to a decline in coffee consumption from the eighteenth-century golden age of the coffeehouse.[39]

The history of the coffeehouse in the seventeenth and eighteenth centuries illustrates how public space in its inception, and with its specific quality of publicity, was always mediated by the power relationships characteristic of the emerging capitalist system. Coffeehouses were cosmopolitan and internationally intertwined, but this did not preclude local power arrangements conditioning the nature of public spaces and shaping the effects they would come to have on the fortunes of the public domain. Accordingly, while Habermas may have been right to point to the importance of spaces of consumption in the emergence of a cultural identity for the bourgeoisie based around conversation and the exchange of news, as his critics have argued, it does not follow that these spaces of bourgeois engagement conceived of themselves as forums of democratic reason or that they necessarily promoted a politics of democratic emancipation. Equally, they were sites for the articulation of an undemocratic liberalism, for the sharing of anti-Semitic and ethnic nationalist sentiments, and, in England, for the expression of the cultural superiority of a Whiggish elite that had no intention of bringing the common people into politics and public life.[40] Elsewhere, the café did become a site for the development of quite different kinds of public: a proletarian culture in nineteenth-century Paris and, in fin-de-siècle Vienna, a predominantly Jewish-bourgeois academic and political public sphere that shaped many of the key intellectual trends of the early twentieth century.[41] The transformation of spaces of sociable consumption into the common location of a public is therefore not a matter of the form of those spaces, even if it may be connected to the exact expression of practices of sociability as they relate to the political concerns of the actors involved.

THE BIRTH OF THE RESTAURANT

In the course of modernity, then, both the pub and the café can be seen as sites of an attenuated publicity, with the state and commercial forces serving to hollow out the uniquely public character of free association—with all the political dangers that poses—and to transform them into spaces of mere consumption. A depoliticized sociability becomes an object of commercial exploitation. The process is an uneven one, and, as noted above, the political and public character of the pub and café have been rediscovered and reasserted on several occasions. But it would seem that from its birth, another site of public consumption, the restaurant, is configured so as to minimize public engagement in the performance it oversees. Whereas the pub and the café appear as stages for the public self-disclosure of the patrons, the restaurant is a spectacle with a nonparticipating audience—the diners who sit at separate tables and (usually) do not speak to those at other tables, but only to the intimates with whom they eat. The role of the restaurant patron is to watch the performance of service to which they are subject and to express their approval of the food provided. The restaurant presents itself as a domain of aesthetic contemplation rather than public performance, one in which the food itself now appears as a work of art (or craft) to be assessed on those terms. The price of the food and drink, in that sense, is irrelevant, as too is the matter of whether it lends itself to a conversation of equals. It manifestly does not, as the restaurant operates on the basis of a hierarchical division of the actors involved and the confinement of the nonspeaking audience to fixed points of passive observation at the table.

At least, that is, this is (or was) the common view of the restaurant as it has been given to us in what is said to be its first manifestation: the French restaurant as it emerged in late eighteenth-century Paris. This kind of restaurant provided an extensive list of dishes advertised in a menu, usually also served wine with the food, seated diners at separate tables, often had an ostentatious setting, and delivered dishes with theatricality in service. As Rebecca Spang has shown, this kind of restaurant did not—against much of the established wisdom—appear after the French Revolution—when the cooks of the court and the aristocracy, now dismissed from service, started to prepare food for the public—but rather some time prior to it.[42] For Spang, the key to understanding the institution at its birth is to consider it not as an ostentatious space of theatrical performance (this would come later), but rather in terms of the kind of food it first produced and the setting out of the physical space in which it was eaten. The term *restaurant* didn't refer to the establishment itself, but to the dish it served: a refined bouillon or a kind of consommé composed of long-cooked meat that was supposed to have restorative qualities. The "semi-private, semi-public" space, as Spang refers to it, of the restaurant developed to cater for what were peculiarly bourgeois culinary preferences in the eighteenth century for lone or private dining, with a simultaneous concern for the link between diet and health but also for the display of refinement in taste, which we also encounter in our discussion of distinction in chapter 9.[43] The restaurant offered to the bourgeois customer an alternative to dining out at the public *table d'hôte*—an unpredictable affair, both in terms of the food served and the moral standing of the company one was required to eat with at the shared table. In contrast, as Spang claims, the significance of the restaurant was its promise, as proclaimed on the signs outside the establishment: "I will restore you." It was this "restoration for individual eaters" that constituted the appeal of the restaurant at its birth.[44]

In the nineteenth century, the Parisian gastronomic culture of *la grande cuisine* was promoted by food essayists and chefs of public renown, most famously Jean Anthelme Brillat-Savarin and Marie-Antoine Carême.[45] After the revolution, this movement also represented the beginning of the idea of distinct national cuisines and cooking styles, with French cooking being widely regarded as the most superior and sophisticated form of national cuisine, much sought after by the bourgeoisie across Europe (see chapter 8). The fashion for eating out in restaurants did not, therefore, take off in London until the French chef Auguste Escoffier and the Swiss hotelier César Ritz set up in the Savoy. At this point, the commercial imperative driving the expansion of restaurant numbers is clear to see. As Rachel Rich claims, we do not encounter the restaurant here as "a response to a crisis in aristocratic life," but rather as "a cultural artefact of a society bent on commodifying all aspects of leisure and sociability."[46] Eating out in the restaurant might be considered as a form of display, the parading of wealth and cultural capital. But as Frank Trentmann has recently argued, the modern world of consumption is driven not primarily through the emulation and imitation of one's "betters" but by the inculcation of habits and routines that are shared with one's peers.[47] The growth of the restaurant as a commercial space interspersed widely within the fabric of the city is a necessary condition of the practice of going out to eat, a practice so diverse in the modern world as to bring into question the usefulness of the category of a "restaurant" that corresponds to the ideals of *la grande cuisine,* opulent display, and theatricality. Unquestionably, these features continue to inform ideas of "fine dining," though even the bastion of French gastronomy, the *Guide Michelin,* has in recent times been forced to recognize the role of contemporary forms of informality, the importance of ethnic cuisine, and even street food in the dining scene.[48]

The variety of restaurants that have developed since the nineteenth century raises once again the question of the relationship between sociability and publicity in such spaces of consumption. What is notable about the restaurant in its nineteenth-century French manifestation is that it sustains only a limited sociability of the diners at the table and positively discourages the kind of interaction between guests at different tables that may lend itself to civic or political engagement. The forms of publicity that involve performative conversation mediated by customs of sociable drinking and eating are absent in the restaurant, even though they may have been present in eating spaces that preceded it. As Beat Kümin argues convincingly, while the Parisian restaurant undoubtedly involved the innovation of individuals eating out for leisure, rather than for sating hunger or engaging in ritual and civic acts of collective eating, there was a great deal of continuity between eating in public in early-modern Europe, particularly in the inns and taverns, and eating in the restaurant. Inns and taverns often catered, on request, for individuals or small groups dining in privacy, and there was a wider selection of dishes available, not just those prepared on the day by the host.[49] Yet forms of restaurant emerged in nineteenth-century France that maintain a number of the innovations of the "classic" model, but which—quite consciously—reinvest both the food served and the space of association for diners with public sociability. The *bistrot* and the *brasserie* appealed to different class audiences and, in the case of the former, to a local sociability that might cut across class divides. Thus, the various forms that the restaurant has taken since its birth in the eighteenth century follow the pattern we have noted of other public spaces of consumption. They can lend themselves to commercial imperatives that involve the exploitation of both workers and customers and can reinforce class divisions and other social distinctions, but at the same time, they can be sites for a form of public sociability that lends itself to processes of political and social democratization.[50]

CONCLUSION: EATING OUT AND THE PUBLIC-
PRIVATE DISTINCTION TODAY

If we are concerned with the connection between spaces of consumption and social and political change, we may want to consider what is signified in various practices of eating out. Eating outside the home is a practice characteristic of (if not confined to) the modern world. It is a product of the long-term changes to the food system we have noted in previous chapters, but it is also a variety of consciously and deliberately followed activities that have themselves come to shape the modern food system. This two-way relationship between practices of eating outside the home and the food system clearly impacts on the relationship between the public and the private that was discussed at the beginning of the chapter. The distinction between the public and private is not straightforwardly "social," mapping neatly onto the division between the space of the household and the arena of civil conduct outside the home. It is rather a political distinction, in the sense that what defines the home and its outside is often determined not only by the formal legal rules and procedures of political authority but also by the outcomes of political struggles between collective actors, including the black civil rights sit-in campaigns mentioned in the introduction. While eating out today thus may seem like a set of practices that are confined to the enjoyment of public sociability, we cannot extrude such practices from the political context in which they take place and the political impact they have on the organization of the contemporary food system.

Recent decades may be seen as marking a growing democratization of eating out and a gradual undermining of the class basis of eating distinctions characteristic of the nineteenth and much of the twentieth centuries.[51] The greater access of individuals to food knowledge—including knowledge that was once the preserve of the gastronome—made available through the growth of guide books and then the internet,[52] has been accompanied by a growing informality of eating styles—a process that started in the home in the twentieth century and spilled out into forms of public eating and drinking. The dining scene in cities across the world shows considerable diversity, with food available in a wide range of public venues—from food markets to food courts in shopping malls and sports stadiums, as well as to cafés and restaurants in commercial outlets, not least supermarkets.[53] The decline of the family meal, which has supposedly accompanied mass eating out, has long been a worry for conservatives, as has been the transformation of gender roles within the nuclear family with the reentry of women into the workforce. There are, however, good grounds for thinking that recent changes in social relations within the private domain of the family do not mark the atrophy of the home as a site of consumption but rather its transformation.[54] Developments in food and related technologies—such as refrigeration and electric lighting—made mass eating out in the twentieth century possible, but more recent innovations in transport and communications technology have allowed takeout to transform the space of domestic consumption.[55] While takeout was for a long time associated with poor quality and unhealthy food and the undoing of familial sociability (as it was often eaten straight out of the box in front of the television), it has become increasingly acceptable—and, indeed, indicative of refinement if one orders at the right place—to feed guests at home, even at middle-class dinner parties, with food that has been ordered online and then delivered or collected.

There is a sense, then, in which eating out has become eating in, and vice versa. In the last few decades, it has become commonplace to claim that the public domain is under threat, with a decline of the "social capital" that comes from face-to-face sociability in public spaces,[56] or by the appropriation and closing down of public space by private corporations and other agencies in the era of neoliberal capitalism.[57] Notwithstanding these concerns, it should be evident that there has been a proliferation and diversification of public spaces of consumption that offer the possibility of the development of new forms of social capital and the reclaiming of public space from private and corporate commercial interests. The consumption of commodities, including food, is not simply the vehicle for the advancement of an unfettered capitalism, for, as Arjun Appadurai has argued, it is the social relationship that we have with things that determines their value for us.[58] At the same time, the commodity's value is not simply socially mediated within a system of exchange, but is politically determined in much the same way as is the public-private distinction. Eating out today cannot be simply a matter of experiencing the pleasures of sociability; it is already implicated in a food system that is politically constituted, and eating out—and increasingly eating in—can be seen as political acts. These acts may serve to reproduce the food system, but their very nature and reach mean they also have the potential, when consumers themselves think and act politically, to transform it and pose a challenge to the contemporary political and economic order.

CHAPTER 7

The Modern State

Alcohol, Alcoholism, and Biopolitics

On first viewing, William Hogarth's famous prints *Beer Street* and *Gin Lane,* issued in 1751, present us with starkly contrasting judgments of the value of two different alcoholic beverages. *Beer Street* seems to be a celebration of the life of the stout English wright, typified by the black-smith on the left, a figure made hearty by a lifetime's consumption of beer and beef. The denizens of Beer Street enjoy all the comforts of an honest life of labor and good cheer. The verse in the subscript tells us:

> We quaff thy balmy Juice with Glee
> And Water leave to France.
> Genius of Health, thy grateful Taste,
> Rivals the Cup of Jove,
> And warms each English generous Breast
> With Liberty and Love.

In contrast, the wretched inhabitants of Gin Lane suffer the terrible afflic-tions brought on by addiction to a foreign drink.[1] Business booms for the pawnbroker, Mr. Gripe, as the dead are buried in the background, a body hangs from the rafters of a derelict building, and, to the fore, a prostitute casts her child over the side of the stairs leading to the gin shop in favor of a box of snuff. Debauchery is all around in this, the realm of the idle:

> Gin cursed Fiend with Fury fraught,
> Makes human Race a Prey;
> It enters by a deadly Draught,
> And steals our Life away.

FIGURE 5. *Beer Street,* by William Hogarth, 1751. *"Beer Street" by William Hogarth (British, London 1697–1764) via The Metropolitan Museum of Art is licensed under CC0 1.0.*

Hogarth's illustrations may be taken as a paean to the virtues of work and restrained joy, an exaltation of the tradition of English liberty and a condemnation of the permissive lives of the indolent poor in thrall to a foreign power. But as usual in Hogarth's work, there is more going on than at first meets the eye. The residents of Beer Street seem oblivious to the suffering of their fellow citizens, and the biggest villains on

GIN LANE.

FIGURE 6. *Gin Lane,* by William Hogarth, 1751. *"Gin Lane" by William Hogarth (British, London 1697–1764) via The Metropolitan Museum of Art is licensed under CC0 1.0.*

Gin Lane are not the impoverished drunks, but those who enrich themselves—the pawnbroker and distiller—by preying on the vulnerability of the poor.

The prints appeared as England was at the height of the "gin craze," an episode we shall consider in more depth presently. Hogarth's work speaks to the concerns of this chapter in a way that is not immediately obvious, for not only is he addressing the social conditions of the poor—

an activity that will be of increasing political importance in the wake of the Industrial Revolution—but also, albeit obliquely, he is raising the question of the relationship between the state and the everyday lives of the people who live under its authority. In Hogarth's day—and this continues to be the case in many quarters of political and legal theory—the relationship between the state and individuals was conceived of in largely legalistic terms. The state, as a sovereign entity and the supreme power in a given territory, through its laws provides the conditions in which its subjects may live in civil peace and freedom (or at least a freedom consistent with the freedom of all others) and, in return, expects obedience to those laws on pain of punishment or even death. In classical liberalism, the state's functions are confined to the constraint of those internal enemies who threaten civil peace and the vanquishing of external enemies who threaten the state's existence.[2]

Today, this legalistic picture seems far away from a reality in which the state's reach into society is far more extensive than was conceivable in the eighteenth century. The state and its agencies play a central role in regulating and prohibiting things and ways of living in order to meet a variety of goals. In the state's regulation of (and, at times, prohibition of) the production and consumption of alcohol over the last three centuries, there have been a number of factors at work: the promotion of national wealth and well-being, the maintenance of social order and conformity, and the satisfaction of religious-cum-moral views of propriety in conduct. But in pursuing these ends, a host of institutions and bodies of specialist knowledge formed outside the state have been involved: medicine and the medical profession, churches and other charitable and voluntary organizations, workhouses and schools, and the social sciences and social services. The French social theorist Michel Foucault (1926–84) used the terms *biopower* and the related *biopolitics* to describe, in general, the relationship between these disparate bodies and practices and the state in the government of the life of populations as a whole.[3] From around the seventeenth century in western Europe, the body of the population comes to be seen as a form of life, a life that must be constantly extended and reproduced. In this regard, the line between the state and society, as well as between private and public life, the drawing of which is a central feature of liberal political thought, becomes blurred.

In this chapter, we will focus on the production, distribution, and consumption of alcohol to disclose important features of modernity and, in particular, the relationship between state and society. As we will see, a concern for the ways in which alcohol has been used and abused over the last three centuries demonstrates some key problems of state-society relations in the modern world. To illustrate these problems, we will turn to two episodes in which the state has tried to control alcohol production and consumption: the gin craze in London (which resulted in the Gin Acts of 1736 and 1751) and Prohibition in the United States between 1920 and 1933. First, however, we will set the scene by considering alcohol in human history.

ALCOHOL IN HISTORY

Practices of drinking alcohol mark forms of social identity and social distinction, uniting and dividing people in ways that relate to economic class, gender, power, status, and divergent cultural tastes. Drinking is not a biological function, but a wide series of socially conditioned practices. The only common material element of these disparate practices is alcohol. In chemistry, *alcohol* refers to a number of different chemical compounds, some of which, like methanol, are fatal to humans even in small amounts. What we commonly call alcohol is one particular type of chemical, ethanol. The human body can break down and process ethanol effectively in small doses, but in larger amounts, it is a toxin causing a range of short-term to long-term physical and psychological symptoms (including relaxation, euphoria, nausea, paranoia, depression, chronic disease, and death). The content of ethanol in alcoholic drinks varies very widely. Historically, most beers have contained relatively small amounts, being usually less than 4 percent alcohol by volume (ABV).[4] Wine has also been weaker in the past than it tends to be today, usually around 7–12 percent ABV, though modern winemakers can produce wines that exceed 15 percent ABV (for reasons outlined below, they usually cannot produce anything in excess of around 16–17 percent ABV without fortification). Spirits tend to be around 35–50 percent ABV, though some very strong spirits can go over 90 percent ABV.[5]

Ethanol is produced by fermentation, a process that occurs naturally. Yeasts on the surface of a fruit react with the sugars it contains when it ripens and the skin is permeated, producing carbon dioxide and alcohol. It is highly likely—though, of course, unverifiable—that humans and their hominid ancestors first happened on the pleasant physical effects of consuming alcohol by eating fruit that had begun to ferment naturally. In the wild, there are instances of animals seeking out and ingesting alcohol and other intoxicants.[6] The history of the origins of deliberately getting drunk must remain speculative, but it would seem probable that at some point (or rather at numerous different points), fruits were gathered with the intention of allowing them to ferment and eating or drinking the result. The first concrete evidence we have of people making alcoholic drinks is from the Neolithic age, roughly 9000 years ago. The Neolithic, as discussed in chapter 2, saw the flowering of agriculture and the birth of the first cities. While it is probable that the first alcoholic drinks to be purposefully produced were made from grapes, dates, or other soft fruits, and possibly from watered-down honey (mead), the cultivation of cereals would have lent itself to brewing, and traces of fermented grains and seeds have been found in clay pots used at the time.[7]

The production of fermented drinks has then been going on for thousands of years, for the most part on the domestic scale.[8] Drinking alcohol has been both a mundane and a numinous activity over this period. For many agricultural laborers, the drinking of beer was a routine part of the daily order, not least because it was often far safer to drink than water—the fermentation process kills off many of the pathogens present in contaminated water. At the same time, the consumption of alcohol has been a central feature of magical divination and religious ritual. Among other narcotics, alcohol has been used by shamans to induce trance-like states. The drinking of wine, most notably in the Christian Eucharist, symbolizes communion with the divine, a religious practice that has its origins in the cult of Dionysus in ancient Greece and Bacchus in Rome. Wine has often been seen as the preserve of the aristocracy and priestly castes, a mark of social distinction from the beer-drinking common people. Forms of drinking—and forms of drink—have thus been centrally involved in the constitution of social order for millennia, but they have also been involved in social disorder.[9] The containment of that disorder can, in part, explain the permission of the extraordinary, irreverent, and excessive in the festival, the most obvious incarnation of which is the carnival.[10]

Concerns about the effects of alcohol consumption on social order are very old and a key factor in religious prohibitions on drinking. For example, take the Islamic proscription of alcohol, usually said to be derived from these verses of the Qur'an: "You who believe, wine and gambling, idolatrous practices, and [divining with] arrows are repugnant acts—Satan's doing—shun them so you may prosper. With wine and gambling, Satan seeks only to incite enmity and hatred among you, and to stop you remembering God and prayer. Will you not give them up?" (5:90–91). As Sami Zubaida argues, unlike the consumption of carrion, pork, and blood, which are forbidden on the grounds of their impurity in the sight of God, alcohol and gambling are prohibited on the grounds of social order; as Satan's work, they distract the people from their religious and mundane obligations.[11] The social character of the forbiddance has allowed elites to justify their drinking (of wine in particular) on the basis that it does not inhibit their performance of office or the observation of rite. Historically, the enforcement of the prohibition has, however, involved the political authorities, as when from time to time in the Ottoman Empire, the taverns (and coffeehouses) were closed down due to the political dangers posed by uncontrolled public assembly.[12]

Age-old concerns about the connections between intoxication and disorderly conduct are amplified in modern history by two developments: the distillation of alcoholic spirits and industrialization. Distillation allows for the concentration of alcohol in volumes that could not possibly be achieved by fermentation. The reason that wine and beer cannot be normally brewed beyond around 16–17 percent ABV (though some less common strains of yeast can tolerate slightly higher levels), is that the alcohol present in the must or wort kills off the yeast and thus halts the fermentation process. The only way to increase the ABV further is to drive off the water content of the fermented solution by heating: as alcohol has a lower boiling point than water, its vapor can be captured and condensed, thus resulting in a much higher level of ABV. Distillation as a process was likely in use by the fourth century c.e. by Greek alchemists, though there is no indication that they were distilling alcohol at this point rather than other substances such as mercury. The evidence points toward alcohol first being distilled by the Arabs around the eighth century, and the etymology of *alcohol* would seem to support the notion (Arabic: *al-kuhl*), as does the name of the vessel in which distillation took place, the *alembic* (Arabic: *al-anbiq*).[13] In Europe, the alembic was used to distil alcohol from the early fourteenth century, but until the sixteenth century, it was used first and foremost for medicinal purposes.[14] "Spirits" were usually seen as having restorative and vital effects, reflected in their description in various languages as the "water of life": *aqua vita, eau-de-vie, akvavit, uisce beatha,* and *uisge beatha.*

But the discovery of alcohol distillation does not by itself account for the modern concerns with excessive drinking in society. The production of alcoholic drinks was taken out of the domestic sphere and placed in the hands of specialist brewers and distillers capable of producing alcohol in very large quantities. The same processes we identified in chapter 5 that led to fundamental changes in the food system in industrialized societies also apply to the production and consumption of alcohol. Even prior to the Industrial Revolution, brewing and distillation had become commercial activities on a significant scale.[15] It is not just an expansion of the supply-side of alcohol that takes place in the early modern period, but also a transformation of the spaces and culture of drinking. If drinking had largely been a private matter, confined to the home or a routine part of the workday for agricultural laborers in the fields, it became an increasingly public affair in early modern Europe (see chapter 6). For drinking to become a problem for the state, it had to be visible to it in the state's own domain: the public space. Over time, however, we can see how that public space comes to be displaced by "society," and drinking comes to be regarded as a social problem. It is now neither a private nor a public matter, but a threat to society that must be contained or eliminated.

FROM SOVEREIGNTY TO BIOPOLITICS: THE GIN CRAZE

There is general agreement in the literature of political and historical sociology that by the mid-eighteenth century, the contours of the modern state in northern and western Europe had been set. The modern state emerges as the *sovereign territorial state*, a form of political organization in which there is a single, undisputed source of law that applies to a well-defined territory and that is backed up by the state's control of the means of violence within that territory.[16] This kind of state appeared in Europe largely as a result of monarchs defeating their internal enemies and, in doing so, gaining monopoly control both of the means of wealth extraction (taxation) in their territory and of the armed forces that could be deployed against other states in the pursuit of their dynastic, extraterritorial ambitions.[17] Among varied sources of taxation, the consumer habits of the state's subjects provided particularly rich pickings. This was evidently the case in a city like London, a major center for the movement of goods and people in which a large working population had access to a wide variety of consumables. The Glorious Revolution of 1688—the deposition of James II by the combined forces of Parliament and the Stadtholder of the Dutch Republic, William of Orange—marked the beginning of a new armed struggle as England entered the Grand Alliance against French expansionism in Europe. The

British state geared up for war and sought out new tax resources to fund its campaigns.

There were two economic consequences of the war with France that would promote the widespread drinking of gin. First, a ban was placed on the import of French brandy, which up to this point had been the favored spirit drunk by the working population of London. Second, sitting on a surplus of grain, large farmers—many of whose landlords occupied the benches of Parliament—sought a new market for their grain. In 1690, "An Act for encouraging the distilling of brandy and spirits" was passed by Parliament, and there was an immediate spike in the production of hard liquor. The monopoly of the Company of Distillers in London was lifted, and anyone could distill without a license.[18] Significantly, they were also free to vend what they produced, and a large number of outlets distilling and selling spirits appeared in London in the early 1700s; one calculation is that there were 8,579 "brandy"-shops by 1736, but a large number of gin sellers were located in the poorest parts of the city, mainly Southwark and the East End around Whitechapel.[19]

Gin—the name for which was derived from *geneva,* not a reference to the Swiss city, but rather to the Dutch spirit *genever,* named after its flavoring with juniper berries—became the favored drink of the poor by the 1720s, supplementing (rather than replacing) strong beer, which continued to be drunk in large amounts.[20] While estimates vary,[21] it is clear that gin consumption increased significantly through the 1720s. Some historians, echoing the moral concerns of the time, have claimed that this represented "an orgy of spirit drinking" by the indigent and threatened the (self-)destruction of "civilized life in Britain."[22] But as Peter Clark argues, these views tend to exaggerate the scale and effects of the spirit trade, much of which was carried out by existing victuallers.[23] What occurred between 1720 and 1751 in London, in many respects, looks like a series of moral panics[24] as campaigners, led by prominent figures in arts and letters, such as Henry Fielding, Daniel Defoe, and—as already noted—Hogarth, in combination with magistrates and lawyers associated with the Society for Promoting Christian Knowledge, such as Sir John Gonson and Sir Joseph Jekyll, applied pressure on Parliament to limit and prohibit the production and consumption of gin.[25]

The first wave of the movement against gin came in the wake of the economic downturn that accompanied the bursting of the South Sea Bubble in 1720,[26] but the campaign reached its peak later, first in 1735–36 and again in 1750–51. In 1736, the campaigners succeeded in seeing the first Gin Act passed by Parliament, which sought to all but ban the production and sale of spirits. It raised taxes on spirits, introduced a prohibitive annual license for vendors of £50, and provided for the payment of a reward for informants revealing unlicensed sellers to the authorities. This was supposed to stop the small-scale distillers and hawkers, who had been identified as the main agents of gin's spread. In fact, it had little effect on production and consumption and actually prompted greater social disorder by fueling violence against the magistrates attempting to implement the act and the informers they used. Seeking revenues to fund Britain's participation in the War of the Austrian Succession, the act was superseded in 1743 by a new piece of legislation that slashed the license fee while further raising excise on spirits. This signaled the end of the crackdown of 1736.[27] The campaigners pressed on, however, and in 1751, on the back of the great popular success of Hogarth's *Gin Lane*, a new prohibitive act was introduced, increasing the excise on spirits by more than 50 percent, banning the sale of gin in prisons, and outlawing back-street distilleries and the hawkers.[28]

While the movement against gin might look, on the face of it, like a religiously inspired campaign against sin, of the kind that was prominent in Christian Europe throughout the Middle Ages in times of economic insecurity, war, and civil conflict, it exhibited fundamentally different concerns and strategies for achieving its objectives. No doubt religious belief played an important role in motivating individual campaigners, but the general tenor of the movement pointed to the deleterious effects of gin consumption on the health of society as a whole and on economic life, law, social order, and public health more specifically. Mid-eighteenth-century London saw the development of concern for a range of social problems, from public squalor to the fate of abandoned children. Foremost among these concerns with respect to the consumption of gin was its effect on women and the family. It is no coincidence that the figure placed at the center of Hogarth's *Gin Lane* is a "fallen woman," supposedly representative of the many women who had migrated into London from the countryside only to find themselves cast into prostitution and drunkenness, shamelessly and openly sharing the company of men in the alehouses and gin shops.[29] The state was increasingly seen as having a responsibility for the consequences of this corruption of women's role in the family. The purpose of the state's provision of the conditions of economic prosperity came to be seen as its commitment to deal with these public—and eventually "social"—issues that private citizens could not see to alone.[30] But private citizens did deal with these social concerns in a more systemic manner through new charities and foundations. "Charity" moved beyond the provision of alms and took on an institutional character, witnessed, for example, in the establishment of the Foundling Hospital in Hatton Garden.[31]

More generally, public problems were ever more viewed through the lens of science and medicine. The gin craze was regarded, in Fielding's terms, as "a new kind of drunkenness," and writers such as Thomas Wilson and Stephen Hales were quick to declare it a disease. This was not just a disease of individuals, but of the social body as a whole, evident in the physical degeneracy of infants born to gin-addled parents. The spread of the disease threatened the economic and political greatness of the nation.[32] But the idea was to be shown to be more than an intimation through the marshaling of statistical evidence. As Warner puts it, the moral reformers were engaged in a calculation of political arithmetic by drawing on official and semiofficial statistics to make their arguments. This involved using a wide range of figures on levels of wealth and its relationship to population—rates of fertility and infant mortality—and statistical evidence about the increase in the volume of gin produced and sold, in order for the reformers to press home their case.[33]

The gin craze can be located in the sixty-year period from 1690 to 1750, and in Foucault's work, this corresponds to a time in which we see the emergence of new strategies or technologies for the government of people's conduct.[34] The 1690 Act embodies sovereign power, where *sovereignty* is understood as the power "exercised over the land and the produce of land" and "concerns power's displacement and appropriation not of time and labor, but of goods and wealth."[35] The theory of sovereignty—or, as Foucault calls it, "the juridical theory of sovereignty"—grants legal title to the sovereign to extract goods and wealth from its territory (principally in the form of taxation). A law to encourage distilling and increase the amount of revenue garnered from excise on spirits fits well into this model. But Foucault claims that over the course of the eighteenth century, two other forms of power emerged that, in important respects, superseded—if not eliminated—the operation of sovereign power. The first of these was *disciplinary power,* which "is one of bourgeois society's great inventions . . . one of the basic tools for the establishment of industrial capitalism and the corresponding type of society."[36] In his book *Discipline and Punish,* Foucault sets out how this disciplinary power is concerned with the production of "docile" bodies that "operate as one wishes, with the techniques, the speed and the efficiency that one determines."[37] These bodies are disciplined to make them subject to the requirements of industrial capitalism: workers who are habituated to repetitive and monotonous labor. This form of disciplinary power was exercised in institutional settings where the individual could be constantly scrutinized and made to conform to a highly routinized set of practices governing bodily conduct. These institutions were the prisons, workhouses, hospitals, and schools into which many indigent drunks and their offspring were placed. Behind the "humanitarian" concerns of the moral reformers who sought to banish gin stood the incipient doctrine of utilitarianism, coded most clearly in Jeremy Bentham's scheme for the Panopticon, the first total institution in which, with minimum resources, those who presented the greatest challenge to social order could be constantly surveyed and made into willing contributors to the *summum bonum*.[38]

It is, however, the other form of power that Foucault identified as appearing in the eighteenth century that does most to illuminate the gin craze and subsequent attempts to prohibit the production and consumption of strong spirits. Biopolitics is "the attempt, starting from the eighteenth century, to rationalize the problems posed to governmental practice by phenomena characteristic of a set of living beings forming a population: health, hygiene, birthrate, life expectancy, race," a form of power that governs the population as a whole.[39] The concerns of the anti-gin movement were, centrally, concerns about the health of the population, measured by a growing body of statistics regarding public health. The state would come to adopt those concerns, being at the forefront of attempts to reduce infant mortality, prolong life, and inhibit the degeneracy of the "race." Excessive drinking of spirits was regarded now as a disease that corrupted the social body, turning men and women away from their social obligations: for men to work hard and for women to bear and raise healthy children who would reproduce the nation. Drinking, in this respect, was increasingly regarded not as a public activity—to be seen and celebrated in the shared space of alehouse or the tavern, as on Hogarth's *Beer Street*—but as a social problem, an ever-present danger to society that must be controlled and even eliminated. *Gin Lane*, then, was not just a representation of the problems of a neighborhood, but a picture of what would happen to the social body when degeneracy and disease were invited in—an imaginary that has arguably been remobilized through the "obesity epidemic" discussed in chapter 11.

DRINKING OUT OF SIGHT: PROHIBITION

Between 1920 and 1933, Prohibition in the United States saw a nation-wide ban on the production, sale, and transportation of all "intoxicating liquors." The ban was given constitutional force through the Eighteenth Amendment to the Constitution. The National Prohibition Act of 1919 (better known as the Volstead Act), which set out the federal laws governing the ban, defined an "intoxicating liquor" as any beverage containing more than 0.5 percent ABV. Yet as well as permitting exceptions, such as the use of wine in religious rites and the production of alcohol for industrial use, the act allowed individuals to continue to consume in private whatever stock of drinks they had in their personal possession, and it went even further by exempting home-brewed cider and other "fruit juices" that were not subject to the 0.5 percent limit, but only prohibited if a jury were to determine that they were "intoxicating in fact." The latter exemption was designed to appease that part of the rural population who had most vehemently supported the campaign for prohibition, yet wanted to continue to be allowed to drink in the privacy of their homes.[40] Prior to Prohibition, consumption of alcohol in the United States had been far from a marginal activity. During the eighteenth century, most alcoholic drinks were in the form of distilled spirits, and of these, rum formed the main part. In the early nineteenth century, spirit consumption rose considerably, to over five gallons per capita per annum. At the same time, spirit drinking was supplemented by the consumption of fermented drinks: beer, wine, and hard cider (about 10 percent ABV). In this period, Americans were drinking more than the English, Irish, and Prussians, and about the same as the Scots and the French.[41]

Prohibition, then, did not seek out the elimination of drinking, but rather of all forms of visible public drinking. In important respects, this was not a departure from but a deepening of drinking practices that had developed in the United States since colonization. In New England, the Puritan colonists, while not forbidding drinking—and, indeed, in many respects singing the praises of a "moderate" intake—nevertheless sought to limit the number of taverns.[42] In his famous work *Democracy in America*, Alexis de Tocqueville noted how the Connecticut code of laws of 1650 based penal legislation on scripture, the legislators' primary concern being "the preservation of moral and good practices in their society; thus they proceed continually to penetrate to the heart of man's conscience and not a single wrongdoing escapes the condemnation of the magistrate." The code stipulates that "laziness and drunkenness are harshly punished. Innkeepers cannot serve more than a certain quantity of wine to each customer." Yet Tocqueville resists denouncing these laws as despotic, as they were part of Puritan democracy, "freely voted in by all the interested parties, whose customs were even more austere and puritanical than the laws." The injunction against public drunkenness, as for injunctions against using tobacco and even sporting long hair, was to protect the congregation as a body of equals who freely decided on their laws.[43] While public houses, particularly in Boston, would go on in the late eighteenth century to be distinctively political spaces, in which opposition to and eventual revolution against the British Crown would be formed, elsewhere, as in Philadelphia, the taverns often took on a socially and politically exclusive character and appeared not as part of a public domain, but more as quasi-private spaces in which men with similar religious, class, and ultimately political affiliations would associate.[44]

The main organized agent of Prohibition in the late nineteenth and early twentieth centuries was the Anti-Saloon League.[45] For members of the league, support for abstinence (rather than moderation) was first and foremost a matter of religious calling, but in the pursuit of Prohibition, there was a relentless focus on the mobilization of its resources to achieve this single goal through political and constitutional means.[46] It established a formidable organization that applied pressure on elected representatives to support the temperance cause and to act as a counterweight to the lobbying power of commercial brewers and distillers. At the same time, it became allied to the cause of the "Progressives," Protestants from the northeast, such as Theodore Roosevelt, who saw the saloon as the engine of corruption that moved the machine politics of the large cities of the East and Midwest. Much of this view of the corruption of the saloon was shaped by anti-immigrant and anti-Catholic sentiments, but also out of a commitment to social and political improvement.[47]

The temperance movement, as it developed during the nineteenth century, has been seen as an organization of the rural Protestant middle class, an essentially antimodernist reaction of small-town America to an increasingly urban and industrial society and to newer waves of immigration, primarily Catholic.[48] Joseph Gusfield argues that the cause of achieving Prohibition through political mobilization can be seen as a "symbolic crusade," an effort to defend the status and cultural superiority of Anglo-Saxon, middle-class Protestantism.[49] From another perspective, the temperance movement might be regarded not as the last gasp of a social group in demise, but rather as the positive expression of a process of individualization characteristic of the modern world. In Norman Clarke's words, "This developing consciousness of the individual, rather than communal dignity, this turning inward for new sources of individual direction, destiny, and discipline . . . was coming to dominate American and European life during the nineteenth century." Sobriety, on this view, was a mark of personal responsibility to lifestyles based around the values of the nuclear family: conjugality and domesticity. Accordingly, "The movement gained power in American society as that society rejected the older, more open, even public style of life and began the internalization of its loyalties, energies, sentiments, and disciplines."[50]

The problem with Clarke's view is that by the 1920s, when Prohibition was legally in force, practices of drinking were emerging that were precisely concerned with the expression of individuality within the nuclear family. To drink did not involve an abdication of responsibility, but rather formed part of a lifestyle centered on—the ideal at least—of individual fulfillment through the nuclear family and in the domestic sphere. It would take some time after Prohibition for this lifestyle to become mainstream, but with the advent of mass advertising of branded drinks through television in the 1950s, drinking at home in the presence of the family became widely accepted as normal and desirable. At the same time, Prohibition served to accelerate the development of a kind of permissive individualism that started in the illegal bars and nightclubs—the famed speakeasies—of major cities such as New York and Chicago and that would spread much further afield and be articulated in emerging forms of mass media: the radio, the gramophone, and motion pictures.[51]

One key effect of Prohibition, then, was not to end practices of drinking, but rather to accelerate their displacement into the private sphere of intimate relationships. At the same time, public spaces of alcohol consumption became increasingly oriented around intimate relationships, concerned less with the sociability and conviviality of transient meetings between strangers in public spaces and more with the formation and enforcement of emotional and sexual bonds. But while this private, quasi-public space of drinking became established and accepted by middle-class America in spite of Prohibition, Lisa McGirr has shown that in another respect, the impact of Prohibition proved far more enduring. The enforcement of Prohibition targeted particular sections of the population above all others: the poor white working class, African Americans, and urban immigrants. A new federal system for governing crime came into place, under the auspices of the Federal Bureau of Investigation and other agencies, that constituted a new penal state. Despite the failure of these agencies to control the production, transportation, and sale of alcohol during Prohibition,[52] they were the model for the first war on a prohibited substance, a model that has been deepened and expanded in the ongoing war on drugs, but which also provided the grounds for other campaigns concerned with crime and national security.[53] Thus Prohibition contributed to the genesis of forms of self-government not by enforcing the discipline of abstinence, but by mediating alcohol consumption through intimate relationships and eroding further the public sphere of sociable drinking. Yet Prohibition was also founded in the interstices of sovereignty and biopolitics. The combination of the state's power over the life and death of its subjects in the field of territorial security and the concern for promoting the life of the social body—manifested in measures to eliminate social degeneracy and disease—had and continues to have a profound effect on social relations in the present.

CONCLUSION: ALCOHOL AND SOCIAL PATHOLOGY

Intoxication and *drunkenness* designate two different conditions. Intoxication is a physiological state that can be measured by markers such as blood alcohol content or breath alcohol content. These measures can inform clinical and legal definitions of intoxication, as, for example, in the stipulation of drunk driving limits. Due to the dissimilar ways in which different people metabolize alcohol, it is not uncommon for people to be technically intoxicated without appearing drunk or to appear drunk without being intoxicated. The reason is that drunkenness is a quality of visible conduct rather than of chemical balance. The drunk comports himself or herself in particular ways, though precisely what counts as drunken conduct varies very widely. Drunkenness is often a shared social activity that possesses its own ritual practices and regulatory norms. There is thus frequently an expectation that the celebration or marking of certain life events—the birth of a baby, marriage, the mourning of the dead—warrants getting drunk. Other occasions, such as religious festivals, national holidays, and sporting events, are also often attended by drunkenness. Equally, there are judgments concerning inappropriate drunkenness in these settings, but these are not related to levels of physiological intoxication, but to forms of conduct that generally—regardless of whether the person performing them is intoxicated—invite disapprobation.

We have seen in this chapter that episodes of the state's prohibition of alcoholic drinks in modern societies depict drunkenness not as a socially regulated practice but as a pathology of the individual and social body. The word for this pathology, *alcoholism,* carries with it the idea of the drunk as addict, constantly desiring the physiological state of intoxication. Here, drunkenness is collapsed into intoxication. As Maria Valverde claims, from the late eighteenth century, habitual drinking came to be seen not simply "as a bad habit but as a disease, a 'palsy of the will' [in the words of the American physician Benjamin Rush]."[54] The drunk is not responsible for his behavior, but is driven to it by his addiction. In 1849, a Swedish doctor, Magnus Huss, published a book with the title *Alcoholismus Chronicus, or Chronic Alcoholic Illness.* As Jean-Charles Sournia shows, Huss not only coined the term *alcoholism* and charted the pathophysiology of long-term drinking but also was active in the Swedish temperance movement. He argued that alcoholism was not just an illness of the individual, but of society as a whole, because of its degenerative effects.[55]

Huss's work played an important role in the medicalization of what was regarded as excessive drinking, particularly the drinking of spirits. But while addiction was initially associated only with such heavy drinking, the consumption of other narcotic substances and eventually other forms of behavior such as gambling or sexual promiscuity came to also be seen as instances of addiction. In the mid-twentieth century, this more general understanding of addiction was established, in one definition, given by the United Nations Commission on Narcotics in 1950, as "a state of periodic or chronic intoxication, detrimental to the individual and society." The same commission proposed a distinction between "addiction-producing drugs," such as opiates and cocaine, and "habit-forming drugs," such as analgesics and amphetamines, which were "not generally considered to be detrimental to the individual and society."[56] The habitual use of a variety of drugs thought to be inherently addictive was considered an illness, whereas the consumption of others—such as tobacco—was not. Yet the "behavioral addictions"—to gambling, sex, or eating—point not toward the inherent addictiveness of those activities, but to the "addictive" or "compulsive" personality.

Habitual patterns of consumption or conduct unquestionably have a physiological aspect. But the construal of alcoholism as an organic disease of the individual and social body obscures the constitution of widespread contemporary concerns with the effects on health and social order of "addictive" behavior. A historical perspective on addiction and its manifestation in events such as the gin craze and Prohibition show that whatever its physiological character, it only becomes a matter of concern when it comes to be regarded as a social problem. The current war on drugs recapitulates many of the constitutive tensions in the relationship between state and society that we see in these earlier episodes. As an economic agent, the state's pursuit of significant revenues from tax on alcohol (as well as tobacco and other legal drugs) may conflict with its and other quasi-public agencies' commitments to maintaining the health of the population as a whole. The state's policing function—its coercive control of crime and the enforcement of disciplinary mechanisms to promote social order and economic efficiency—runs into demands for individual self-realization through consumption and claims that the mastery of addiction is a matter of self-government rather than paternalistic control.[57] Even if we are witnessing the decline of state sovereignty today—though there are many reasons to believe that this is a much more contestable and uneven process than usually thought—the widespread view of alcohol consumption and drunkenness as a social problem continues to underpin the state's role in its government.

CHAPTER 8

Identity

Nationalism, Ethnicity, and Religion

The first step in the affinity between people and food lies in ecology. For most of human history, most people ate what was available in the lands where they lived, typically a limited and seasonal range of what could be collected, hunted, and later planted, as discussed in chapter 2. We saw in that chapter and in chapter 3 that variety and complexity of food cultures are made possible by cities and trade, often as parts of large political units, such as empires. Following the second Industrial Revolution, this diversity was enhanced by the rapid transportation of railways and steam ships, as explored in chapter 5. This complexity, however, is often restricted to the affluent classes, while the majority of cultivators and workers remain confined to a limited range of staples. It is in the context of these possibilities that distinctive cultures develop, associated with social identities of city, class, ethnicity, and religion, but always subject to economic and cultural transformations, such as colonial expansion, the breakup of empires, and, in more recent centuries, the upheavals effected by the spread of capitalism and world markets. It is in these more recent contexts, with the formation of territorial nation-states on the ruins of broken empires, that the idea of "national cuisine" takes root and is associated with ideologies and historical constructions of nationalism. As such, they become possible subjects of ideological contentions regarding "ownership" of cuisines between national and communal borders, drawing out the binaries of self and other, nature and society, and the eminently modern processes of eco-

FIGURE 7. Chicken Balti Pasty. Ginsters of Cornwall.

nomic globalization, social stratification, and political transformation outlined in the introduction. The existence of a variety of rival and contested national food cultures is a key condition for the formation of personal identities in the modern world. But, as we shall see in this chapter, contemporary processes of globalization are reinforcing ideas of distinct ethno-nationalist food cultures while at the same time eroding the institutions and practices of ethno-national and religious membership.

THE AGE OF NATIONS AND NATIONALISM

The nineteenth century, following the French Revolution and then the Napoleonic wars, saw the rise of a new kind of politics and state formation: nation-states and nationalist ideologies.[1] The French Revolution itself comprised the ideology of nationalism and common national belonging and citizenship. The diverse populations and languages of France—Celtic, Germanic, Occitane, and Basque—were all pressed into this ideological unity, and the processes of capitalist modernity that followed facilitated this linking of diverse elements.[2] Napoleonic military conscription and then the spread of education and literacy in the French language facilitated the slow and partial integration of the diverse elements. The ideas of nationalism and citizenship spread through Europe and subsequently other parts of the world, creating what has been called "reactive nationalism."[3] The later nineteenth century saw the unification of both Germany and Italy, with further nationalist ideas of ethnic/racial and territorial roots of culture and ethos. These models and ideas, in turn, had great influence on the nationalisms of countries in central and eastern Europe and further east into Eurasia, notably Japan. The breakup of empires following the fortunes of World War I—the Habsburg Austro-Hungarian, the Russian, and the Ottoman—led to the creation of other putative nation-states in Europe and the Middle East, and yet more were formed with the wave of decolonization of the European empires in Asia and Africa following World War II. The nation-state became the formal model of political unit in the world, enshrined in the United Nations and its institutions. It is obvious, however, that the entities thus created are widely diverse, and the idea of national character and unity is often contested within and between states. Ethnic and religious diversity, amicable or antagonistic, are regular features.

The ideas of ethnic and communal cultures within nations are part of the contentions. "Culture" is the basic ingredient of nationalist ideologies: the idea that the identity of the ethnic or territorial group that becomes the nation is preserved through history by attributes of language, custom, religion, and traditions. The claimed unity and continuity over time are often constructed histories, the "invention of tradition."[4] Food is an important element of these cultures and has the added importance of connecting the people to the soil, the *terroir*. The peasant and rural life and customs are often romanticized in nationalist narratives, naturalized as the basic ingredient of the nation. While premodern aristocratic and bourgeois attitudes to the peasant were typically contemptuous (see chapter 9, "Distinction"), and this continues to be the case in everyday parlance, the peasant is raised in nationalist discourse to a noble figure in the constitution of the nation and its natural relation to the soil.[5] Food nationalism is accentuated in the case of the breakup of old empires and the formation of nation-states, each claiming the heritage of the previously cosmopolitan empire. As the modernist or instrumentalist theories of nations and nationalism indicate, cultural practices related to language, the arts, history, and landscape are standardized and codified by the state and nationalist ideologues in an effort to unify otherwise dispersed and diverse expressions of identity.[6] Food is one powerful medium in both the construction of ostensibly national or ethnic culinary repertoires, and in the everyday reproduction of a "banal" nationalism by marketing, labeling, representing, and regulating certain foodstuffs and beverages as national or ethnic products.[7] The challenges of doing so are conveyed by the image of the Chicken Balti Pasty reproduced above, which manages to combine in one food item references to the British nation, the Cornish region, and the city of Birmingham, where, in the

1990s, "balti houses" reinvented a dish from the northwestern Pakistani area of Baltistan, traditionally cooked in metal vessels alternatively called *karahi* or *balti* ("bucket" in Urdu, Hindi, Odia, and Bengali).

The French was the first cuisine with a national label to be celebrated and imitated in Europe (as we will see in the next chapter). From the seventeenth century it was a courtly and aristocratic cuisine that was followed all over the continent, and French cooks were much in demand. This supremacy and fame persisted into later ages and, after the Revolution, into bourgeois milieus and eventually into restaurants and media, albeit transformed at every junction. In the age of nationalism, rival national cuisines were declared or assumed, notably in Italy after its unification in the latter part of the nineteenth century. Food cultures in Italy had always been (and remain, in certain respects) regional, and as John Dickie demonstrates, tied to the major cities and their elites. The rural classes and the urban workers, like their French equivalents, had limited and poor diets. It was after unification that an Italian cuisine was proclaimed in the landmark publication of Pellegrino Artusi's *Science in the Kitchen and the Art of Eating Well,* in 1891.[8] This became the most influential and constant reference in Italy on matters gastronomic; Artusi introduced Italians to their food and established a repertoire that became Italian national cuisine. "Pellegrino Artusi not only brought together and codified many of the recipes that are still the mainstays of the peninsula's cuisine, but he also turned those dishes, for the first time, into a template of Italian national identity."[9] What are now common items, such as pizza, were for the most part restricted to particular regions (pizza to Naples, for example) and scarcely known elsewhere in the country—and when they were known in other regions, it was often in peculiar versions. Artusi's pizza, for instance, was an almond pastry.[10] The spread of pizza in Italy coincided with its popularity in the wider world, especially from the 1970s, and the center for this spread was the United States. We shall turn presently to the role of diasporas in shaping national cuisines, of which Italy is the prime example.

Artusi's collation of Italian food coincided with socioeconomic and cultural transformations ushered in by modernity. Social and geographical mobility of populations, expansion of cities, literacy, education and media, transport and communication, and military conscription all brought people from different regions into close interaction, a gradual process of national formation, including food. Nationalist reactions against the dominance and prestige of French food led, in many countries, to the proclamation of the virtue of native food. In England, as we shall see in the next chapter, the reaction, early on, was articulated to social and religious contests between the French- and Catholic-affiliated monarchy of the Restoration (1660) and a bourgeois Protestant opposition. The dominance of French cuisine as elite and prestige food continued until recent times, equally denounced by more puritanical and nationalist sentiments: elaborate sauces, cream, and garlic allegedly used to disguise bad ingredients.

THE GLOBAL STAGE AND THE NATIONAL IDENTIFICATION OF CUISINE

A dominant theme in the discourses on globalization is the homogenization and "McDonaldization" of foods, with the spread of American mass culture and fast food. Witness the spread of McDonald's and other hamburger chains, pizza parlors, hot dogs, Coca-Cola, corn flakes, and much else of the same ilk to the corners of the earth, at the expense of local, national, and regional foods. This expansion is, indeed, perceived as a threat by nationalist defenders, a notable example being the movement against McDonald's in France in 1999. José Bové, an activist and charismatic peasant leader in southwest France and a producer of Roquefort cheese, led a group of followers to dismantle a McDonald's restaurant under construction in the southern town of Millau.[11] He emerged as a national hero in France and to the anti-globalization movement elsewhere. He denounced an agricultural model that sees food purely as an industrial commodity rather than the centerpiece of national culture and family life.[12] These arguments are similar to the sentiments and ideology behind the Slow Food movement, which originated in Italy in the 1980s and spread its chapters and producers to many parts of the world.[13]

Fast food and U.S. cultural imperialism, however, are only one aspect of globalization. Globalization also generates an opposite trend: a global stage on which food identities—national, ethnic, and regional—are declared, elaborated, and commercialized. In effect, people are being asked: what is your national/ethnic cuisine? And in response, they narrate and invent that cuisine.[14] Trade, travel, media, and diasporas feed this process. The ever-growing public interest in food, now a recreation and a spectacle, amplifies this process. Innovations and fusions often retain identity labels in combination, such as Moroccan hummus, unheard of in Morocco. Food nationalism, then, is heightened, if not created, by this staging.

The example of the formation of Italian cuisine in the United States is a good illustration of diasporic formation of national cuisine. Between

1876 and World War I, it is estimated that seven and a half million Italians migrated to the Americas, primarily to the United States, but also to Argentina, Brazil, and other parts of South America. They were primarily peasants and workers driven to migration by poverty and hunger.[15] Large numbers from various parts of Italy settled in American cities, creating "little Italies," where many engaged in the food trades of kitchens and groceries. They were particularly attached to their foods as markers of identity in their new country, but they also indulged in the new abundance as against the dearth and hunger from which they fled, able to eat meat, mortadella and salami, and much pasta in rich sauces. Families from the various regions and diverse food cultures of Italy were neighbors and members of common associations, and they intermarried, in the process creating a fusion cuisine of Italian food, which they then presented to the wider American society. Popular dishes were invented that had no precedent, such as spaghetti with meatballs.[16] Through the twentieth century, many of these dishes became part of world cuisine, notably the now ubiquitous pizza. Thus, Italian national cuisine, as it is presented to the world, was synthesized to a considerable extent in the American diaspora. This is a notable and well-documented example of the diasporic formation of national cuisine. We shall consider further examples in the discussion of Indian cuisine below.

THE MIDDLE EAST

The Middle East region (Western Asia, with echoes in North Africa) offers good illustrations of historical trajectories of food cultures and the nationalist and ethnic claims on that history in the modern world. The region, as we briefly saw in the chapter 2, is home to ancient agrarian civilizations, including the Babylonian, Assyrian, Hittite, and Persian. It was then brought under the rule of successive and rival empires: the Hellenistic, Roman, later Persian, Byzantine, Islamic, and, most recently, the Ottoman, which left the deepest traces. Each empire contained syntheses of the diverse cultures it inherited and dominated, with wide regional variations, and this hybridity was fertile ground for innovation, facilitated by military, commercial, and cultural interchanges. The Abbasid Empire, for instance, often considered the golden age of Medieval Arab culture, drew heavily on the Persian heritage it ruled, with elements of the far simpler background of Arab ancestry. Aristocratic cooks and connoisseurs bequeathed manuscripts of recipes and banquets, repeated and imitated in subsequent periods.[17] Many of the recipes bear Arabized Persian titles, such as *sikbaj*, a stew of meat in vinegar and honey, deriving its name from the Persian word for vinegar, *sirke*. Ingredients and methods traveled to the Arab kingdoms of Spain and ultimately became incorporated in the culinary cultures of the country, where hispanized Arabo-Persian appellations survive to the present time, such as *escabeche*, from *al-sikbaj*, but now typically designating fish pickled in vinegar. In turn (as we saw in chapter 3), *escabeche* became the South American *ceviche*, marinated raw fish. The dish and the label became largely extinct in the Middle East. This example illustrates the complex syntheses and migrations of food culture-epitomized by the Columbian exchange—following successive imperial syntheses.

The Abbasid Empire was ended by the Mongol conquests of the thirteenth century; its once brilliant capital of Baghdad was sacked and destroyed in 1258. The Mongol "hordes" have a bad reputation in evaluative history as destroyers of civilizations, from China to various Muslim lands, with little of their own nomadic culture to contribute, beyond horse meat and fermented milk. Yet the Mongol conquests played an important role in connecting and bringing together elements of the civilizations they conquered and into which they were eventually assimilated. Rice, dumplings, and tea were instances of such transmission from China into Central Asia and Persia. Turkish peoples, equally nomadic and "tribal," were in due course to invade and dominate the Middle East, the Balkans, and further afield, where they were to establish one of the most brilliant and long-lasting empires, the Ottoman (1298–1922), with a series of syntheses that are most pertinent to the region in modern times.

The Ottoman dynasty, starting in 1298, ruled over patches of the Balkans and Anatolia at the expense of the Byzantine Empire and the Seljuk state (another Turkish dynasty, 1077–1308). It conquered Constantinople in 1453, then reached an apogee of territorial expansion in the subsequent century, controlling the Arab Middle East, North Africa, the Balkans, much of southeastern Europe, the Caucasus, and parts of Iran. It was one of the longest-lived empires in world history, finally ending only after defeat in World War I in 1922, after a long period of decline after the increasing encroachment and control of the ascendant European powers, beginning in the eighteenth century. The Ottoman centuries effected long and complex cultural syntheses in language, religion, science, law and statecraft, music and art, and food and drink. The major elements were the Byzantine heritage, the Arabo-Persian of previous Muslim empires and lands, the Turko-Mongolian military and statecraft, and elements from the other territories of Europe and the Caucasus. These were dynamic and evolving syntheses, with many regional and social variations, with innovations and imports, but at the same time, the Ottoman territory included vast areas of isolated and static rural and tribal enclaves, of mountain and desert. Many of its regions bordered the Mediterranean and thus participated in the exchanges and influences around that basin of cultures. The regions bordering the Black Sea shared in different cultural perspectives of Russia, eastern Europe, and the Caucasus.

Modern nation-states, with various degrees of cohesion and viability, arose on the previous Ottoman territories, each comprising a mélange of ethnicities, religions, and cultures, but with "national projects" of integration, with diverse nationalist ideologies. Pan-Arabism, for instance, competed with nationalism, such as Egyptian and Iraqi. These latter drew on historical constructions of their ancient civilizations, the Pharaonic and the Mesopotamian, distinguishing each nation from the more inclusive categories of "Arab" or "Muslim," though acknowledging both. In these ideologies, the Ottoman heritage was presented as an alien intrusion that had held up the progress toward modernity and civilization. Even stronger rejections of the Ottoman past featured in Balkan nationalisms: the Greeks, Bulgarians, and others. Yet that heritage leaves many traces, certainly in the respective culinary cultures. Typically, these traces are "nationalized," declared to be original to the particular nation, appropriated by others. Thus common and desirable items are fought over, as baklava, the well-known sweet pastry, is claimed by Greeks, Turks, and Arabs.[18]

Global processes and events, notably the migration of foods from the Americas after the fifteenth century (see chapter 3), but also the later transformation of the world by the expansion of capitalist markets and the dominance of European powers from the eighteenth century, led to wide-reaching impacts on food cultures in the Middle East. The tomato, the potato, the haricot bean (the much-esteemed *fasoulia* of the region), the pimento, and corn were all incorporated into that food culture, with the tomato becoming an ever-present ingredient in many dishes. These episodes may be ignored by nationalist and ethnic narratives, which insist on the antiquity, originality, and continuity of cultural essences. Few people are aware of the American origins of those common ingredients.

There is a familiar repertoire of Middle Eastern food at the present time, with variations on common themes in different communities and regions. Common items such as kebab, flat bread, stuffed vegetables, distinctive meat and vegetable stews, meat balls and dumplings (kofte and kibbe), and special dips and salads—including the now ubiquitous hummus, transformed beyond recognition in the global sphere. These are part of the evolution of the Ottoman synthesis and of related Iranian repertoires. To say "Ottoman," however, is not to say "Turkish," as it is the product, as we have seen, of a mélange of cultures and histories. "Ottoman cuisine" is now used in the global lexicon as a label of prestige, connoting refined courtly cuisine, and this is certainly embraced by Turkish cooks and entrepreneurs. Yet, as we will see in chapter 9, "Distinction," Ottoman courtly cuisine was refined but limited—and, in any case, the product of diverse cultural diffusions. Much of what developed in the different regions of the empire was the inputs from various regions and communities, many of which had their own distinct adaptation and idiosyncrasy. This diversity within a common repertoire gives wide scope for identity claims of nationality and ethnicity.

To illustrate the above points, let us consider the particular food region comprising southern Anatolia (Turkey), the Arab Levant (Syria/Lebanon/Palestine), Cyprus, and into Northern Iraq and Western Iran. Until the ethnic cleansing and massacres of the late nineteenth century and into World War I, the population was an ethnic mix: Turkish, Kurdish, Arab (to the present), Armenian, Greek, Assyrian, Jewish, and other minority ethnic religions.[19] This distinctive food repertoire includes many of the items that have become globally familiar through trade and migrations and are now known in the food lexicon as "Eastern Mediterranean," such as hummus, tabbouleh, and other salads and dips; kofte and kibbe; pastries savory and sweet, such as the range of *boureks* and baklava; and *lahmajun/ lahm-b-ajin,* spiced ground meat on flatbread crusts, now part of the fast-food repertoire in Europe as "Turkish pizza." There are, of course, variations, within and between particular cities rated for their specialties—notably Aleppo and Gaziantep—and particular communities, such as Kurdish dairy products of their mountain pasturage. Following the drawing of national boundaries in the postwar settlements, the massacre and displacements of the Armenians and the population "exchange" of the Greeks, this common heritage was claimed by each nationality/ethnicity as its own. The Armenians, renowned in the region as fine cooks and caterers, carried this repertoire to their diasporas as "Armenian."[20] Armenians played an important part, historically, in connecting Iran to this region. Between the sixteenth and early nineteenth century, they were the main agents in the silk trade between Isfahan and Aleppo, the latter being an entrepôt to the Mediterranean and Europe.[21] On the culinary front, we note the presence in Aleppo's cuisine of ingredients characteristic of northern Iran and the Caucasus, such as pomegranate and walnuts, in dishes such as *Muham-*

mara salad, and pomegranate molasses as flavoring for *lahm b-ajin,* salad dressing, and olives.

The Kurds, disqualified as a nation by the post–World War I divisions, are ever anxious to assert their cultural identity, including the national claim over the culinary heritage. This heritage has spread to the global food stage through Lebanese, Cypriot, and Turkish eateries and groceries in many parts of the world, including the West, claimed variously by the particular national cuisine. In Turkey, this Anatolian food came to Istanbul and other major cities with Anatolian migrations, coming to prominence from the 1980s in regional restaurants and kebab grills, which are called Gaziantep, after the city, which is renowned for its fine food. The Istanbul bourgeoisie were condescending to a cuisine they considered inferior to their fine traditional foods. They denounced the spices and strong tastes as crude and "Arab"—much like the English disdain for European food—in favor of their plain, undisguised ingredients flavored only with fresh herbs, and they preferred their refined rice pilafs over the bulgur (cracked wheat) of Anatolia. Fast-food lahmajun saloons were symbolic of this alien invasion, much like the distinctive Arabesque music, both reflecting a sociocultural process of othering. Turkey, of course, comprises other distinct regional foods, crossing other boundaries. The Pontic (Black Sea) region and the northeast of Anatolia border the Caucasus and includes ingredients and recipes common in neighboring Georgia and Armenia. It is also renowned throughout the country for its love of *hamsi,* a kind of anchovy, which is featured in many dishes, including a pilaf and a bread made from corn, equally typical.

Now, on the global stage, we have "Turkish Cuisine," at times elevated to "Ottoman Cuisine." These labels are in cookbooks,[22] restaurants, and travel and media sources, and they are, of course, celebrated and defended as part of a national heritage with historical depth. We have seen, however, how it was formed in stages of cultural syntheses in a multiethnic urban setting of a cosmopolitan empire. Regional diversity and overlap of national borders had also meant that southern Anatolian food came to northern and western cities as a novelty but also became involved in cultural contests and regional-ethnic stereotypes. On that global stage, this national designation faces challenges from other ethnicities and nations, both in their own countries and in diasporas and in international markets, as demonstrated in the common food repertoire of the geographical location of southern Anatolia and its Arab neighbors, with a diversity of ethnicities. Similar challenges are posed for other national cuisines in the region.

In the Arab world, cuisines are diverse and regional. We have seen that the Arab Levant shares a distinctive repertoire with southern Anatolia and Cyprus. And it is this repertoire that had become prominent on the world stage as Lebanese food, thanks to the spread of Lebanese restaurants in the rest of the Arab world and in the West. Syrian and Lebanese cities share with Istanbul and the Aegean cities a history of public eating and tavern culture, unknown or less prevalent in other Arab lands, except for Egypt. This is partly due to the enterprise of Christian communities in these regions (Armenian and Greek in Turkey) who manufactured and served alcoholic drinks and the food, meze, that accompanied drink.

This tavern culture translated, in more recent times, into restaurants. In the rest of the region, public eating was confined for the most part to cook shops and market stalls, mostly fast food and takeout. The Arabian peninsula and the Persian Gulf had, historically, limited ingredients and cookery, being, for the most part, desert with a nomadic people. The exception were the cities of the gulf coast that were, and remain, cosmopolitan enclaves, fed by trade—and later oil—with different ethnic settlers: Iranian, Indian, East African, and later many more nationalities constituting much of the working populations of the oil cities. The other exception were Mecca and Medina, which, as centers of the annual pilgrimage and historically having populations from all over the Muslim world, engaged in religious pursuits and trade, bringing a variety of foods—Indian, Persian, and Levantine.[23] The cities of Egypt and North Africa were penetrated, from the early nineteenth century, by colonial governments and European settlers. The Egyptian elites, for instance, had been of Ottoman culture but were then attracted to European styles in colonial times.[24] Egypt had contributed important items to the repertoire of Middle Eastern food, most notably *foul medames*, the fava bean stew now common as a meze item, and it laid a claim to falafel, the fava and chickpea rissole that is now a fast food item in many places, most notably Israel, to which we shall now turn.

Iconic Arab/Palestinian foods, primarily hummus and falafel, have been widely adopted by Israelis and are now claimed as Israeli, to the chagrin of Palestinians, who consider the appropriation of their culinary cultural heritage as one more affront after the occupation of much of their land and the continuing encroachment on what remains, as well as their legal and political marginalization.[25] And it is not only Palestinians who are incensed by Israeli claims. The Lebanese consider hummus iconic to their cuisine, and falafel is claimed to have an ancient heritage in Egypt, where it is also known as *ta'miyya*. *Hummus* means "chickpeas" in Arabic, and the full name of the popular dip, *hummus b'tahina*, also refers to the sesame paste that goes into the preparation. This dish is original to the Arab Levant and Anatolia. Until the middle of the twentieth century, it was little known elsewhere in the region. Israelis point to the Jews of the Levant, such as those of Aleppo, as originators of the dish and, as such, claim a title for common or cosmopolitan "ownership." Though hummus and much else from the region are now commonly seen on menus and supermarket shelves globally, the question of "ownership" still rankles in the ideological disputations of the region.

INDIA

India is a fascinating example of a nation-state composed of an acknowledged diversity of regions, ethnicities, languages, religions, and cultures. It is also the heir and product of diverse empires, each adding its peculiar synthesis with native Indian cultures. India's food cultures are further complicated by the rituals of caste, specifying limits and taboos applying differentially by caste status.[26] Examining the syntheses and determinants of "Indian cuisine" is particularly illuminating to the issues of this chapter.

The Muslim Mughal Empire (1526–1857) shaped many aspects of regional and ethnic cultures, including food. It introduced meat cookery to a largely vegetarian milieu of elite India dominated by high castes. Its interactions and exchanges with the Turko-Persian world of Iran, Afghanistan, and Central Asia brought many ingredients and methods into the courts and elite circles, some filtering down to urban classes. Delhi, Agra, Hyderabad, and Lucknow all comprised courtly societies and urban cultures colored by these elements. Dishes now familiar as Indian repertoire, such as korma, biryani, and tandoor grills, were part of this legacy.

The Portuguese colonized parts of coastal western India, notably Goa, and mixed with the native population, creating various hybrid syntheses. Most notably, they introduced the chili pepper, an import from the New World, into Indian cookery, which slowly spread into other parts of India by the eighteenth century and became iconic to Indian cuisine as we know it. They also introduced pork, previously unknown or taboo for Hindus and Muslims. It was part of a wave of conversion to Catholicism. Vindaloo, now another emblematic item of the Indian repertoire on the global stage, was a Portuguese introduction—the original and "authentic" dish being pork in vinegar and garlic, not the ultra-hot heroic curry of Indian restaurant fame.[27]

The British Empire, first in the shape of the East India Company (1600–1874), then direct British rule, had the most profound consequences for the country. It was a capitalist empire and, as such, a revolutionary harbinger of modernity, disrupting and transforming traditional rural economies and societies. In its later industrial form, in the nineteenth century, it led to the most profound transformations with modern transport and communications—steam ships, railways, and the telegraph—as well as the institutional modernity of imperial government. All this made a great impact on food cultures. It was the British who invented "curry." The British in India maintained items of their native and European diet, symbols of their distinction from the native populations, but also adopted and adapted many elements of local foods, creating hybrids. *Curry* arose in this milieu as a generic term for a variety of spiced stews and sauces. There is speculation over the origins of the word; it might have come from the Punjabi *cuddy,* a yogurt sauce with lentil dumplings, and the southern *kari,* a thin, spicy sauce, but neither word is used in the generalized sense *curry* came to have. Curry was incorporated into the British repertoire back in the home country, and spawned curry powder as a shortcut to spicing, which became a regular item in British and then world kitchens.[28]

The Indian restaurant was another product of British-Indian interaction and later of globalization. The British, with their unified administration of the country and with the intensification of trade, transport, and communication, as well as the political ideas of nations and constitutions, provided the conditions for the emergence of the idea of India as a nation and country.[29]

THE IMAGINATIONS OF REGIONAL AND NATIONAL CUISINES

Arjun Appadurai presented an interesting analysis of the social interactions in the modern Indian cities that facilitate these imaginations.[30] The grand cities of India—Delhi, Mumbai, Kolkata, and many others—being the centers of government, trade, technology, and culture, are the centers for the new middle classes, the entrepreneurs, professionals, and employees in these spheres, who come from the various regions of the country. These milieus of employment and residence promote contact and sociability between individuals and families from previously separate and even segregated communities, localities, and castes. These educated middle-class groups typically share political and cultural commitment to the idea of India and a national identity. Food plays an important part in these spheres of sociability, in domestic entertaining, markets, and restaurants. People show off their regional foods and discuss the different local versions of common dishes and ingredients, arousing curiosity and experimentation. These explorations also find their way into the media and cookbooks of "regional cuisine." A national cuisine is thus imagined as a generic synthesis of the communal and regional, with common themes and ingredients. Traditional discourses about food in India were primarily about ritual rules and prohibitions, articulated to medical canons on diet and the body, purity, and danger. The social processes of modernity, as described by Appadurai, then generated a different discourse of taste and identity, making it possible to imagine the regional and the national.

Another powerful factor facilitating the identification of an Indian cuisine is the Indian restaurant. Restaurants were largely alien to Indian practice. As we see in chapter 6, the restaurant was a relatively modern creation, developing in France and Europe in the late eighteenth century and spreading to other parts of the world much later. Public eating in many places took place in market stalls, cook shops, and taverns, as well as on collective ritual and communal occasions. Indian caste rules made it more difficult for elite groups to eat in public, inhibited by the many prohibitions and taboos. Low castes and Muslims ate in market and street stalls, but elites considered the cooking of their domestic cooks to be far superior to and more hygienic than public food. The Indian restaurant, then, was a later creation, first found in Britain, then spreading elsewhere, including to India.

One early route by which Indian food entered Britain was through the upper and middle-class India hands—ethnic Britons in India—who carried over their habits and nostalgia into British life. This wave was accompanied by its brand of restaurant.[31] In the early nineteenth century, there were coffeehouses with Indian themes, such as the Hindustanee in central London. The Veeraswamy, the oldest surviving Indian restaurant established in 1936 on London's Regent Street, was also a product of this wave. It was started by Edward Palmer, an India hand and descendant from a family of Anglo-Indian grandees.[32] He imported Indian spices and ingredients, as well as cooks and waiters. The menu featured the dishes of Anglo-Indian classifications—boona, korma, madras, vindaloo, and such like—boasted as authentic. His high-class clientele included the king of Denmark, who is reputed to have sent a barrel of Carlsberg beer to the restaurant every Christmas to ensure the availability of his national beer on his visits. This reinforced the "tradition" of drinking lager with Indian food, now common. Beer was, of course, alien to Indians, but favored by Anglo-Indians from the nineteenth century or earlier, with their imbibing of India Pale Ale with their Indian meals.

Another early route was through Indian students seeking comfort food from home. Some enterprising figures abandoned their studies in favor of providing for this need with restaurants for fellow students and expatriates. Shafi's in London's Soho acted as a community center and club for students and middle-class Indians, and there were a few others. One of the authors of this book (Sami Zubaida) can recall in the 1950s and 1960s the intellectual and student left frequented the Ganges restaurant in Soho, established by a leftist couple, a Bengali and his German wife.

The most important wave, however, was that of Sylheti seamen (Laskars) from today's Bangladesh who abandoned ship and settled, mostly in the East End of London, before spreading out to other parts of Britain. This started in the earlier decades of the twentieth century but increased in the 1940s and after. Many found shelter in crowded boarding houses, often with cafes serving simple curry and rice, started by some of their number with a bit of capital and enterprise. In the aftermath of World War II, there were many humble and derelict properties, old cafes and fish-and-chips shops, offered for sale in different parts of Britain, and some were bought by Sylhetis, who maintained the old workingmen's cafe menu of fish and chips and pies, but added curry and rice. This was the start of the tradition of curry and chips, still prevalent in many cheap cafes. Some of these establishments expanded to full restaurants with Indian food as their main offering, especially in university towns with students looking for tasty and cheap fare. Students typically arrived after a night in the pub, with several pints under their belts. These restaurants spread rapidly to all British towns in the 1960s and became an institution.

Sylhet is a remote part of Bengal, on the Nepalese border, which supplied seamen to and engine stokers for British ships, which took them to British ports and elsewhere. Their traditional cuisine consisted primarily of fish from their abundant rivers, eaten with vegetables and rice, simply spiced. They did not, however, serve their own food (which remains obscure to British and international eaters) in the restaurants. The template of the Sylheti-run restaurant menu was taken from Veeraswamy and the earlier Anglo-Indian restaurants, where many of the

Sylheti restaurateurs had served their apprenticeship as kitchen hands and waiters. Dishes based on meat and chicken, with some Mughal ancestry, became the "Indian food" of Britain and the world, under its own classifications noted above. This food was largely unknown to most of India and unfamiliar even to the meat-eating Muslims and Punjabis. It has, however, now been incorporated in the restaurant culture of India itself. Sylhetis, then, did not inaugurate "Bangladeshi cuisine," but generic Indian. Some restaurants began to claim "Pakistani cuisine," but with no systematic variation from "Indian." It is only in more recent decades in Britain that South Asian enclaves in the main cities have featured their own eateries, often modest and catering to local custom, which were soon discovered and patronized by the wider public. South Indian and Gujarati vegetarian restaurants and *belpouri* houses (Mumbai street food) became popular from the 1970s. Then Punjabi (Lahore) kebab houses, serving meaty cuisines, came to prominence in London and elsewhere, first in the South Asian enclaves of Southall in West London and in the East End, and then spread, in chains, to other parts. There followed Keralan and Goan examples. The globalization of food cultures and the quest for authenticity and novelty since the 1990s has given rise to many Indian restaurants with innovations and diffusions, aspiring to fine dining and Michelin stars, which have actually been awarded to quite a few Indian restaurants in London.[33]

What does the foregoing tell us about Indian food and identity? Indian nationalism and identity are diverse and differentiated. Conservative Hindu nationalists would stress the distinctive rituals and prohibitions of caste and religion. If not purely vegetarian, they would at least denounce the slaughter of cattle, a sacrilege and prohibited by law in some states. On the other side, the cosmopolitan urban classes, as discussed by Appadurai, would celebrate the diversity and synthesis of a national Indian cuisine as developed on the global stage. For the great majority of Indians, however, the question does not arise; the identity of their food is local and familial, bypassing issues of the national and the regional. But, of course, their food habits are being modified by the global trends of production, trade, and cultures of consumption, including the rise of international chains, such as McDonald's and KFC, with local adaptations and cultural value.

India is a unique example of a national cuisine formulated primarily through a succession of foreign empires, elaborated in international locations, and then adopted, more or less, by sections of food professionals and the intelligentsia of the country. Diaspora formulations, however, are much more widespread in the modern world of flux and movement.

THE "OLD NATIONS," FOOD IDENTITIES, AND DIASPORAS

England and Holland, arguably the first venues of capitalism, industry, colonial empires, and modernity, have made only weak claims for a national cuisine and have been receptive to mixes and hybrids of other food styles, including from the colonies, as we saw in the case of Indian food. The English have asserted the superiority of their plain, wholesome food only in defense against the French intrusions and claims of superiority. We will see in chapter 9 how the French developed their claims from the dominance of their courtly styles throughout Europe from the eighteenth century, then continued with the mutations into modern food cultures. Italian cuisine, as we have here seen, was synthesized after the unification of the country in the late nineteenth century and remains highly specific to regions. And what is German cuisine? In all these cases, the constructions of notions of national and regional cuisines have followed similar socioeconomic processes to those advanced by Appadurai for India. France, Italy, and Spain were all originally divided into culturally specific regions, with impoverished and isolated rural majorities.[34] The modern state and the intensification of capitalist markets, with transport (principally railways) and communications, educational systems, literacy, military conscription, the media, and the general intensification of social mobility and public employment, led to the encounters between different communal and regional actors and the synthesis of notions of the national. At the same time, these encounters also facilitated negative or mocking stereotypes of the regional other, such as the mutual ridiculing between champions of pasta and polenta in Italy.

The foremost capitalist countries have been the venues for the construction of the cuisines of other nations, much like the Indian cuisine in the British context. The United States, a country of migrants, is the eminent stage for the syntheses, proclamations, and marketing of national cuisines from elsewhere in the world. There are its old "native" enclaves, such as Louisiana or the Deep South, with their mélange of African and Creole/French elements. New Orleans or Cajun cuisines are then defined and established in cookbooks, media, and restaurants, and, above all, in the marketing of brand names and packaged foods.

Similar processes define other national and regional cuisines, as we saw in the foregoing examples.

THE MEDITERRANEAN DIET

The Mediterranean diet is another identity that was constructed on the global stage, then taken up by actors in the region as a positive designation with marketing opportunities. It was first proclaimed in the 1950s by medical and dietary researchers as a diet supposedly healthier, especially with respect to cardiovascular functioning, than the diets of northern Europe and North America. It was then taken up and developed by food writers and marketing organizations, such as the International Olive Oil Council, and government commercial missions. At the request of a number of countries around the Mediterranean, the Mediterranean diet was declared by UNESCO an "intangible world cultural heritage" in 2013. The constituents of this diet are enumerated as olive oil, as distinct from butter and animal fats, vegetables and fruit, whole grains, legumes, nuts, and fish, with little meat. It also proclaims common social elements of sociability and conviviality around the table, as stated in the UNESCO's proclamation: "Eating together is the foundation of the cultural identity and continuity of communities throughout the Mediterranean basin. It is a moment of social exchange and communication, an affirmation and renewal of family, group or community identity."[35] This supposes that there is some kind of unity of culinary culture in the lands around the Mediterranean basin, brought about by the common ecology and the ingredients it fosters. Once more, the supposedly "natural" determinants of social customs come to the fore in such accounts. The more serious pronouncements often invoke Braudel's seminal book on the history of the Mediterranean. Braudel's characterization, however, is not about uniformity but postulates the sea as a system of war, trade, and cultural exchanges.[36]

The ingredients and cookery of Ottoman lands, including the Mediterranean Arab provinces, pose a challenge to this supposition. Olive oil and wine were not common features (in contrast to antiquity and some European Mediterranean lands). The Ottomans, and most of the Muslim population around that sea, were not very comfortable with fish. Many items of fish in Turkey and Egypt today have Greek or Italian names. In Istanbul, fish is still associated with drinking cultures and most fish restaurants still call themselves *meyhane*, "tavern" and initially had Greek or Armenian proprietors and workers (and some still do). Olive oil was primarily used for lighting and soap and featured on the lists of supply items to palaces and *imarets* (soup kitchens) as such.[37] The preferred vegetable oil was sesame oil. Olive or other vegetable oils were used by Christians during fasts and by Jews who could not use butter (dairy) to cook meat. In many places, such as Egypt and Iraq, they used sesame oil. Wine continued to be produced and taxed in most Ottoman provinces, certainly the Balkans, theoretically by and for non-Muslims. In practice, it was widely drunk by Muslims, not only the elites, but also the soldiery and many among the urban populations. It was not, however, an item of staple consumption the way it was in Latin Europe. It would seem, then, that cultural preferences have considerable weight besides ecological factors in determining patterns of diet; in terms of one of the organizing binaries of this book, society trumps nature. Around the Mediterranean basin, these cultural preferences were diverse, not just between Christians and Muslims, but within each group. It is interesting to note here that olive oil does not appear to have been an item of regular consumption in the European Mediterranean either. Italy, now a major producer and consumer of olive oil, doesn't seem to have used it much in medieval and early modern times; it became common as a food item beginning in the eighteenth century. Even now, Italy and many other European regions use olive oil in limited contexts and continue to cook with butter, lard, and neutral vegetable oils.

There are, of course, discernible commonalities in Mediterranean foods. Some may be determined by the ecology and climate, such as the olive and the grape. However, many common items of culinary cultures were the product of population movements and cultural diffusions. Note, for instance, the role played by Sephardic Jewish migrations from Spain and Portugal to Italy, France, and the Ottoman world, spreading their original cultural items but also adapting new cultures. Seafaring, trade, war, and migrations were other avenues of cultural mixing and diffusion, notably the introduction of New World foods, via Iberia, on Mediterranean axes.

CONCLUSION

To sum up the foregoing analyses and examples regarding food and national, regional, and ethnic identities—historically, "cuisine" as complex combinations of ingredients and styles of cooking was characteristic of urban cultures, typically parts of empires, presiding over diverse regions, ethnicities, and religions and effecting distinct cultural syntheses, aided by extensive trade networks. In this context, food styles were often determined by geography more than by ethnicity. This was illustrated by the example of the Ottoman region of southern Anatolia, the Arab Levant, and Western Iran, inhabited by diverse ethnicities sharing common culinary themes. The breakup of empires into national states, often with ethnic minorities, brought about constructions of national cultures, including the *idea* of national cuisines. These, in some instances, brought about claims and contentions over particular dishes, such as Turkish, Armenian, Arab, and Greek claims that baklava is theirs and the more recent disputes about the Israeli claims to hummus and falafel.

Constructions of national cuisines have a sociological basis in the processes of modernity and capitalism that entail urbanization, common educational curricula, military conscription, common public spheres of media and communications, and social and geographical mobility, bringing people from various regions and ethnicities into common intercourse and participation in urban cultures. Appadurai, as we saw, analyzed this process in relation to Indian cities, as did Artusi before him for Italy after its unification.

Processes of globalization and population movements in more recent times play a prominent role in the national and ethnic identifications of cuisine. We considered the notable examples of the construction of Italian cuisine in the American diaspora and of Indian cuisine in the British imperial circuit. These processes are accentuated in more recent development of cuisine as a sphere of recreation and knowledge, of foodieism, in the global context, through marketing, travel, media, and—as we discuss in chapter 12—the rise of the celebrity chef in all these contexts. These spheres celebrate authenticity, identified in national and ethnic terms, as well as, paradoxically, innovations, which are often framed as the development from national themes. Ironically, cuisine is identified as national and ethnic in the very spheres in which innovation, fusions, and fashion are driving culinary themes in ever more distant directions, resulting in items such as "Moroccan hummus," which must come as a surprise to Moroccans.

CHAPTER 9

Distinction

*Social Difference, Taste, and
the Civilizing Process*

Food cultures in society follow the complexity of the division of labor
and the differentiation of resources, wealth, and power, but an examination of diverse food cultures provides some important insights into the
character of social stratification, one of the key themes of this book. In
terms of social stratification, at the simplest level, difference is by gender
and age, as in some hunter-gatherer groups. Further complexity brings
differences according to wealth, power, and status, including the rituals
of religion and caste. Differentiated and stratified societies have corre-
spondingly differentiated food cultures. A condition for the degree and
complexity of culinary differentiation is the availability of sufficiently
diverse food and drink materials to serve as media of marking difference.
Access to a wide range of materials and to skilled culinary labor under
such conditions follow wealth and status, and culinary knowledge and
sophistication constitute parts of social and cultural capital. Pierre
Bourdieu (1930–2002), the foremost theorist of distinction, postulated
distinction between groups in terms of differential access to two forms of
capital: economic and cultural, a theme that will be elaborated below.[1]
These factors are characteristic of societies that are connected to networks of communication and trade across different ecological and cultural territories, typically, parts of wide-ranging empires, such as those
of China, Rome, or the Ottomans at different points of history.

Jack Goody, in his pioneering work,[2] distinguished two forms of differentiation, which he associated with sub-Saharan Africa and Eurasia

respectively. He considered two types of African societies in northern Ghana, one that had a relatively egalitarian and kinship-based social organization, and the second, in which people were divided by rank and status into chieftains, military commanders, religious scholars, and commoners that in turn, in turn, was differentiated by wealth. They also included slaves (people captured in war). In both social types, however, food materials and cultures showed little differentiation, except in terms of quantities consumed and hospitality offered. The kitchen in both societies was the domain of women, reproducing the domestic economy even under conditions of social hierarchy. This is in contrast to the employment of specialized, skilled male cooks in the high ranks of Eurasian types. Eurasian societies, typically with more complex stratification of class, status, and often ethnicity, and with extensive territorial networks of empire and trade, exhibit more qualitative differentiation of culture and cuisine. Regions that constituted parts of empires, such as the Chinese, the Roman, the Persian, the Ottoman, and, later, the European, were integrated into exchanges, tributes, wars, and trade between diverse geographies and climates and with a wide range of available foods. Peasants and the poor in these societies typically subsisted on a limited range of foods, while the urbanites, the middle classes, the rich, and the nobility had access to an ever-wider range of ingredients and more sophisticated styles of cooking and eating. At the apex was the cuisine of the royal courts, procuring a wide range of food items, including rare luxuries, prepared by hierarchies of cooks, servers, and officers of the household, almost entirely male. Their styles were imitated by associated orders, such as that of the aristocracy, and they, in turn, by the richer merchants and functionaries, and so on. Cities and the urban classes enjoyed greater access to ingredients than the rural populations, typically confined to seasonal products of their lands and subject to the hazards of climate, pestilence, and war. These latter may have conformed more to Goody's African type, with the rich simply eating more of similar ingredients.

Note, however, that differentiation is one not solely of status and affluence but also of various styles of life determined by religion, ritual, and ideology. Indian caste differences present an extreme example of status differentiation by ascetic and ritual styles. The highest caste, the Brahmins, can maintain their ritual status only by wide abstinence: they are strict vegetarians and reject intoxicants. Their food can only be cooked by other Brahmins, usually their wives; their cooking and serving utensils are ritually specified; and they cannot be observed while eating by individuals of lower castes, which limits the company and the venues of their meals, ruling out public eating in restaurants, for instance, and inhibiting travel, considered polluting for their food. This, of course, is an extreme and unique example, but we find elements of it in many complex societies. Consider the religious classes in Christianity and Buddhism: monks engage in frequent fasts and observe ascetic regimes for most of their consumption. Outside the religious and ritual domains, we have many examples of social ideologies and movements that reject luxury and indulgence in the name of higher values and the cultivation of virtue. The great banquets' fine foods and plentiful wine of the Roman rich and aristocracies were censured and rejected by moralists upholding the older republican virtues. Christian and Muslim ascetics and puritans advanced similar criticism and rejection of the indulgences of the rich and powerful.

ROME AND AFTER

The historical trajectory of eating and drinking in the Roman Empire provides a suitable illustration of differentiation. Rome's geographical location in the European Mediterranean provided the basic ingredients of that region. The city was encircled by regions of local production: vineyards and gardens, then wheat, and finally livestock. Wheat, olive, and vine were the three basics, supplemented by vegetables and fruit, fresh and dried, and animal products of meat, fish, and dairy. Animal slaughter was traditionally in the form of sacrifice to the gods, and, as such, meat eating was ritualized. Rome, however, became the center of a vast empire, securing products from much wider regions: half of Europe, the whole Mediterranean and its desert fringes, from North Africa to greater Syria. Northern Europe was known for its carnivorous populations, who imparted more copious and ritual-free meat cultures. The African possessions supplied diverse birds, such as Guinea fowl and ostrich, as well as the exotic fruit of dates. Empire also facilitated trade, with valuable spices from India and Arabia via Egypt. Trade and tribute also brought large supplies of wheat from Egypt and northern Africa, needed for feeding the urban population and the army. And these supplies and imports provided the materials for differentiation and for cultural communication.[3] These materials served as elements of the forms of social difference.

The provisioning of capital cities was a central element of economic policy in all the historical empires—as we saw in chapter 2 and chapter

3—ensuring provisions beyond the operation of markets, by tribute and taxation from the periphery. As such, many Roman citizens (as distinct from provincials and slaves) were entitled to specified rations of food, initially distributions of wheat (later changed to distributions of bread), and later emperors added oil, wine, and pork.[4] On state and religious festivities, guests at banquets received food baskets of different sizes and contents according to their ranks as senators, equestrians, or common people. Such differentiation was also part of the patronage at private banquets, where guests were seated at tables according to their rank and served different kinds of foods and drinks.

The collapse of the Roman Empire was followed by the fragmentation and ruralization of Europe, the contraction and decline of cities that previously thrived on trade. The exceptions were some of the urban centers close to the Mediterranean, mostly in Italy. Much of Europe was "feudalized": kings and princes, lacking resources for centralized administrations and standing armies, contracted military and political functions to feudal lords, who established their rule over rural domains, collecting rents (mostly in kind) and labor dues from tied peasants, and who repaid their sovereign by supplying military support. Under these conditions, life became precarious, both for security and sustenance. War, banditry, droughts, and famine threatened, and trade was sparse and intermittent. The peasant economy and diet was that of subsistence, on the limited foods they produced—the portion that remained after rendering rents and taxes. The nobility had a richer diet, with meat, much of it from hunting game, which they monopolized. However, this diet was simple, lacking the material and sophistication that came with trade and the import of luxury and exotic materials. We see accounts of meals and banquets where great quantities of meat are served: large joints of roasted and boiled meats, poultry of all kinds, including game, with few vegetables.[5] Meat eating was associated with virility and martial power. Hospitality at these banquets was also an avenue of patronage to followers and dependents, reinforcing the power of the lord. The princes of the church dined in similar styles and, while fasting, provisioned their tables with a profusion of allowable items, including various types of fish. The seats of the feudal nobility were typically their castles and estates, while the cities, apart from in Italy, were, therefore, subordinate centers of power and administration, some ruled by a designated bishop and church institutions.

ELIAS AND THE CIVILIZING PROCESS

The precariousness and insecurity under these conditions determine forms of social interaction, conduct, and sentiment, including for food cultures and customs. Norbert Elias (1897–1990) advanced the theory of a "civilizing process" based on examples of the evolution of "civility" in parts of Europe, starting from the baseline of the feudal formations sketched in the foregoing and proceeding through early modern and Renaissance Europe (roughly from the fourteenth or fifteenth centuries, depending on region).[6] Under precarious and unpredictable conditions, people are likely to be impulsive in the expression of emotions and appetites—of fear, aggression, violence, and indulgence of appetite when opportunities occur. The scarcity of food and uncertainty of its availability disposes people to indulge to excess on the occasions of availability, such as feasts and celebration of a good harvest. Rabelais's Gargantua is a caricature of such a man, capable of feats of gluttony so great that they became proverbial. The process of civilization in Elias's theory then proceeds at the level of the political evolution to an ever more centralized state that controls territory and affords a measure of security and the corresponding psychological trend toward greater control and civility. These are also the conditions for population growth, concentration, and enhanced mobility, both spatial and social. The resumption of trade, industry, and urban life in the later Middle Ages affords the resources for centralizing monarchs to institute more elaborate bureaucracies and armies, bringing the feudal aristocracy under control and ultimately bringing them into the courts and the centralized institutions. The resulting courtly aristocracy constitutes a further step in civility, with the development of courtly etiquette, styles, and manners. Part of this process is the greater sophistication of food, banqueting, and table manners. Elias elaborated his idea of a civilizing process by the account of the evolution of table manners.

Elias surveyed manuals of manners, starting in the Middle Ages, through the Renaissance, and up to the eighteenth century, covering the evolution from feudalism to courtly society and the crystallization of the idea of "civility." The instructions are aimed at the elite: the aristocracy and, later, the upper bourgeoisie. The instructions aim at the control of the expression and display of bodily functions, gross appetites, and aggression. The process that Elias postulates is one in which the intensified interaction and the mutual dependence on others for maintaining social status and reputation leads to the internalization of these social controls that, at first, were external demands and constraints.

The tables of the earlier periods were ones of few serving vessels and implements. Food was served on communal vessels: soup in large bowls and meat, in large pieces, carved at table. Wine was often drunk from communal bowls, passed around. Each diner had or shared a knife, often his own, which may also have been used as a weapon, and was provided with a spoon. The fork was to come much later, around the sixteenth century. Diners took morsels of meat with their hands and dipped their spoons and bread into the liquids. Much of the instruction in the manners manuals, then, concerned the etiquette of this communal arrangement: "Do not be the first to take from the dish that is brought in. Leave dipping your fingers in the broth to peasants. Do not poke around in the dish but take the first piece that presents itself."[7] Do not gnaw a bone and put it back in the dish; do not bite a piece of bread then dip it in the broth or the communal cup. There are also injunctions against engaging in bodily expressions at table: do not blow your nose on the tablecloth or wipe your nose with your hand; avoid smacking your lips and snorting; do not poke your fingers into your ears or eyes; do not wipe your hand on your coat.[8] Elias emphasized that all this is said to adults, not children, and that these elementary precepts were given to upper-class people, often with injunctions to distinguish themselves from peasants.

By the sixteenth century, argued Elias, the books of manners related to a court society. The transformation of the aristocracy from the feudal to the courtly entailed the continuous absorption of new elements into the court and the need of members of the elite who were not themselves at court to learn the rules of courtly behavior. This implies that the core of courtly society was formed and its rules internalized, but there was a need for new members and aspirants to be instructed. The preservation of status and reputation required constant attention to the rules and to the pressure and demands of others in the intensified sociality of the court. As more segments of the elites, including the upper bourgeoisie, the intellectuals, and the lawyers, aspired to courtly civility, the books of manners catered to these aspirations.

CITIES

The differentiation between city and country in food culture and opulence was to dominate from the later Middle Ages (the eleventh to the fourteenth centuries) in Italy and continue through Renaissance and early modern Europe. The growth of population, including the urban, and the intensification of trade largely ended the pockets of subsistence agriculture in the countryside and increased pressure on land and labor for enhanced food production, as did the import of certain food items from afar. While the cities were divided by wealth, status, and class, they developed an urban identity, with conscious and proud distinction from the peasants. "The pride that town dwellers felt in their new identity (especially among the lower orders) was such that any risk of being forced to return to rural conditions was seen as a sign of social regression, immediately provoking waves of protest."[9] This civic identity was enshrined in institutions and practices, primarily the craft and trade guilds and corporations, churches and religious fraternities, and, in the case of Italian cities, the aristocratic rulers and church princes. The provisioning of the cities was a function and obligation of those authorities. In the case of Italy, the urban commune, established by those elites, moved to impose authority on the surrounding countryside, becoming, in effect, landlords, feudal masters and rulers, taxing and planning food production to provision the cities.[10] Italian cities, notably Venice, were centers of maritime trade, ranging into many parts of the world, with a variety of commodities, including slaves, but also, of particular interest here, spices. Spices—including cinnamon, ginger, clove, pepper, nutmeg and saffron—were precious and expensive commodities, as was sugar, and, as such, they became the hallmark of culinary opulence and distinction. The markers of status moved away from the mere piling up of meat that was characteristic of the tables of the feudal aristocracy and into complexity and sophistication of cooking employing a wide range of spices and sugar.

While Italy and its cities were the primary locations of flourishing food cultures in those centuries, we should also mention Spain in a different food context. Much of Spain was under Arab rule from the conquest of 711 till the defeat of the Kingdom of Granada in 1492 and the subsequent expulsions of Muslims and Jews. During those centuries, Spain became a conduit for intellectual and cultural currents from the Perso-Arab civilization of the Middle East into Europe, and that included food materials and methods. Rice, sugar, eggplants, and many fruits and spices were introduced and adopted, as well as techniques and recipes. Elaborate sugar cookery—in pastries and worked in eggs and almonds— is widespread in Iberian cookery, some the specialties of convents. Further, Arab translations and adaptations of the philosophical and medical texts of antiquity, including the dominant paradigm of the Galenic theory of the humors, were, in turn, translated into Latin and entered into the European canon. This theory, which dominated medical and culinary practice in much of the Old World till the scientific revolution in the seventeenth and eighteenth centuries, postulated a bodily balance between four humors: blood, black bile, yellow bile, and phlegm. A balance between these elements was required for good health, and this balance had to be managed through diet—the ingredients, in turn, being classified into categories of hot, cold, moist, and dry. Diets and menus were planned accordingly. The culinary legacy of the Arab presence persisted and evolved in Spain after the Reconquista and the expulsions. Urban food cultures of considerable complexity and sophistication characterized Iberian cities, with continuities from Arab rule and into subsequent Spanish and Portuguese centuries, to be enriched further by the imports from the New World (see chapter 3).[11]

John Dickie's *Delizia: The Epic History of Italians and Their Food* chronicles food culture, banquets, and spectacle in the history of the major cities from the late Middle Ages till modern times. Italian cities were the pioneers of urban dominance and culture in an otherwise feudal Europe and were influential in the subsequent evolution of food models on the continent. The city, Dickie argues (uncontroversially) was the locus of inventiveness and sophistication, as against an impoverished peasantry and countryside of limited and monotonous diet. This goes against the common romanticization of the peasant as the origin of simple goodness: peasants produced the food, most of which—and the best of it—was appropriated by the landlords and the cities, leaving them with meager and precarious diets. This image is confirmed in various works on other regions and times,[12] with fluctuations depending on war and peace, population levels, epidemics, and climate. Dickie traces the evolution of food in Italy through portraits of cities, each at a particular point in time, between the late Middle Ages and the modern period. He argues that the food cultures so described pertained to the particular city and period, rather than any general notion of "Italian cuisine," which only emerged with the nineteenth-century unification of the country and in the context of wider nationalist standardizations discussed in chapter 8.

The food styles described show a continuity from the late Middle Ages through the Renaissance—a complexity that uses a wide variety of ingredients, multiple kitchen operations, and strong tastes of many spices, sugar, honey, and vinegar, with an emphasis on display and spectacle. Cooks and savants recorded their creations in a few cookery manuals, the most famous being that of Platina, the pen name of Bartolomeo Sacchi—scholar, priest, and one-time librarian to Pope Sixtus IV in the fifteenth century in Renaissance Rome. His book *De honesta voluptate et valetudine*—*On Respectable Pleasure and Good Health*, is one of the earlier European cookbooks. It records recipes from a veteran chef of a cardinal's household, Martino de Rossi, in a context of theorizing about ethics and health. He summarized the Galenic theory of the humors and its specifications for healthy diet, as well as ethical considerations on the balance between appetite, pleasure, and restraint.[13]

A feast in Ferrara in 1529, celebrating the marriage of the duke of Ferrara's son to the daughter of the king of France, illustrates the excess and theatricality of courtly cooking, as well as its connection to power and patronage. Four courses were served to 104 guests—people of rank, including ambassadors. Each course comprised a dozen dishes, in no discernible order, complex concoctions of poultry, meat, vegetables, and sweetmeats in various combinations, with extravagant use of spices, sugar, honey, and vinegar. Examples that appear weird to modern sensibilities included "sweet pastry tarts deep-filled with the spleens of sea bass, trout, pike and other fish," and flavored with sugar, cinnamon, ginger, cloves, saffron, and pepper. Another is "eel in marzipan."[14] This kind of lavish hospitality was a marker of status and patronage, and a spectacle of power. The large volume of leftovers was, in turn, distributed to layers of servants, dependents, and protégés, serving to cement lines of patronage and power.

Side by side with these lavish displays were the waves of scarcity and famine caused by droughts, wars, and pestilence, affecting the urban poor as well as the countryside. The administered provision for the cities failed under those conditions. In normal times, the rural population and the urban poor ate mostly vegetables, cultivated or foraged, according to season—beans, turnips, onions, leeks, and garlic—as well as rough bread and gruels made from cereals and pulses.[15] Chestnuts were an important staple in areas where the tree grew, cooked as polenta and gruel, especially when waiting for the cereal harvests. Meat was rare and cheese occasional. The elementary equipment and technique, primarily a pot hung over a fire, did not allow for much complexity or sophistication. Vegetables and leaves, the main ingredients of the diet of the poor, were largely shunned by the upper classes, partly in accordance with the dictates of the prevalent humoral theory, which regarded fruit and vegetables as dangerously cold and wet, and as such, a burden on the digestion. They are better cooked, but still too cold and moist. In this model of the body, the stomach was considered like an oven, and digestion as further cooking, and, as such, fruit and vegetables, which are cold and moist, presented a burden. Here, science coincided with considerations of status differentiation from the peasants and the poor. This was to change with the evolution of both medical theories and the "civilizing of appetite," starting in the seventeenth century, primarily in France, and then spreading, in the following centuries, to other parts of Europe.

France in the seventeenth century saw the evolution of courtly and aristocratic cuisine, as well as the emergence of a bourgeois cuisine. These processes were documented with the proliferation of cookbooks and manuals by the prominent cooks of the court and the noble houses.[16] Courtly banquets continued to be lavish, with multiple courses, each consisting of many dishes, but there was a shift in the ingredients and methods: fewer spices and strong tastes, and a new genre of sauces and ragouts. Kitchens were more systematically organized, both in labor and the prior preparation of stocks and sauces. New ways of binding and liaisons—such as flour-and-butter roux, egg-and-fat liaison, thickened meat stocks, and pureed meats and vegetables—partly replaced the old reliance on vinegar, sugar, and honey and thickeners of breadcrumbs and almonds. Vegetables came to play a more prominent part, in line with new ideas about health and digestion. An interesting example is that of green peas, which appeared on the scene at that time and were costly and much celebrated, to the extent of constituting valuable gifts to superiors and mistresses. Green herbs feature more regularly in flavorings and sauces, parallel with less spices.[17]

COURTLY CUISINES AND THEIR LEGACY:
FRANCE AND THE OTTOMAN COURT

France

The French court of Versailles in the reign of Louis XIV (1643–1715) was the height of fashion, much imitated by the royalties and aristocracies of Europe. This is the epoch of centralized absolutist monarchy—the nobility largely incorporated into the royal court in a hierarchy of ranks, engaged in intrigues and maneuvers of power, wealth, and status, all within a façade of civility and elaborate ceremonial. The court was distinguished by continuous spectacles of theatrical and operatic performance and balls, in which the courtiers participated as actors and spectators, and, above all, spectacular dinners, featuring hundreds of dishes, collations, and structures. The king was the most conspicuous diner: he sat at a high table with select members of his family and one or two beautiful courtesans. Other tables were occupied by courtiers according to rank, and the lowest ranks were not diners at all, but spectators. The meal was the spectacle. Those spectators were sometimes let loose after the end of the meal to attack the leftovers in a "pillage," to the amusement of their superiors. We shall recount similar practices in the otherwise different Ottoman court. The king's food was carried in a ceremonial procession, under armed guard, from a dedicated kitchen, part of a vast complex of palace kitchens. While the rules of ceremonial, etiquette, and table manners were elaborate and strict, the food itself was subject to constant invention and innovation by competing celebrity chefs and managers (maîtres d'hôtel). The public spectacle of dining and the tacit competition between the court and the great princely houses reinforced the incentives for innovation.[18] Some of the most prominent chefs of the court and the nobility recorded and published their recipes and techniques, with emphasis on their own innovations and superiority to predecessors and rivals. This publication of recipes continued into modern times, with periodic emphasis on innovation, sophistication and delicacy, against the supposedly heavy cuisine of previous times—there was always a nouvelle cuisine. Another form of publicity was a monthly periodical magazine, Le Mercure Gallant, in the later seventeenth century, which reported on fashion and society, including reports of the grand occasions and banquets, partly a kind of food review, adding to the prominent presence of cuisine in the public realm.[19] The inscription of food in the registers of fashion, sophistication, and innovation, copied in other parts of Europe, introduced a dynamic and a tension in Western culinary culture: the quest for innovation and originality that we see prominently in our own time, but in tension with another quest, one for authenticity and tradition, especially in the national and ethnic branding of food, as explored in the previous chapter.

The Ottoman Court

As it happens, we have an excellent historical study of Ottoman court cuisine from the fifteenth to the seventeenth centuries, based on archives and recipes: Stephan Yerasimos, A la table du Grand Turc.[20] Ottoman bureaucracy was meticulous in recording the quantities of purchases, both for the palace, which fed an enormous number of people, and for the soup kitchens of the capital, which fed students, teachers, mosque functionaries, and designated categories of the poor. Some menus of palace fare at different points in time were also preserved. Yerasimos also resorts to accounts of meals and events by European ambassadors and functionaries. He traces the elements of Ottoman food at these different levels in the classical age of the conquest of Constantinople in 1453 and of Suleiman the Magnificent in the following century. But it is one thing to have records of ingredients and menus, another to know how they are cooked. Yerasimos found the manuscript of a fifteenth-century book of recipes by one Mehmed Chirvani, which was mostly an unacknowledged transcription of the famous thirteenth-century manuscript of Al-Baghda-di's *Kitab al-Tabikh,* but Chirvani seems to have added eighty-two recipes of his own, which Yerasimos surmises were of the fifteenth century. Indeed, many tallied with dishes named in palace menus.[21]

It would seem that the palace meals, including those eaten by the sultan, were, while opulent, relatively simple and repetitive.[22] Two meals a day were eaten, each consisting of two or three dishes. A typical meal: boiled rice with eggs, *manti* (a kind of dumpling or ravioli, in soup or yogurt, probably of Chinese origin, which is still common in Central Asia and Turkey to the present day), and vermicelli in milk. Another meal: *manti* again and *mahallebi,* a blancmange of rice flour and milk with strips of chicken breast. Different forms of *kalye* featured frequently, a ragout of lamb and vegetables or fruit, following the season. The meats were uniformly chicken and mutton or lamb, roasted and stewed. Sugar, honey, vinegar, saffron, cumin, and pepper were regular additions to the soups and stews (again, a common feature in medieval cookery, east and west). Fish was rare (odd, in Istanbul), and when it occurred, it was typically freshwater fish, usually trout from Bursa (though the palace was surrounded by sea on three sides!). The exception to this avoidance of fish was the menu of Mehmed II, Fatih (r. 1451–81), who appears to have enjoyed caviar, oysters, and prawns, all items that were subsequently entirely absent. The medium of cookery was animal fat and copious amounts of butter, shipped from as far away as the Crimea, across the Black Sea. The height of richness and opulence described in poetry was rice dripping with butter. Butter and sugar were the base ingredients for the most luxurious food, pastries and sweets, which were prepared and consumed in great volume and diversity, with extensive kitchens devoted to their production. Drink, in the form of flavored and sugared waters (musk was a most opulent flavoring), sorbets, and syrups, were served after the meal. There is no mention of wine in these accounts, though we know from other sources that wine was commonly taken at court, and many of the sultans were known for their fondness for the stuff. It is likely that the wine table was distinct from the dinner table, as we find in some accounts, including the stories of *A Thousand and One Nights*.[23]

In stark contrast to French court cuisine, in the Ottoman court, while the courtiers ate well, they stuck to particular favorites, homely and repetitive, with little drive to innovation. This contrast is accounted for by the respective sociologies of the court. Mehmed II, Fatih, introduced the practice of the sultan eating alone, often served by pages and eunuchs and sometimes members of the harem. The French kings also dined alone, but in public, with the meal being a display, a spectacle of magnificence and luxury. Cooks and maîtres d'hôtel had every incentive to innovate and shine. Not so with the private and secluded dining of the Ottoman court, which remained repetitive and domestic. There was another important difference of the Ottoman court: it did not feature an independent nobility, peers who participated in ceremonies, balls, festivities, and dinners. The Ottoman court contained dignitaries in the form of viziers (ministers), high clerics, and military chiefs, but the ceremonies in which they participated were highly formalized and stratified. There was a definite hierarchy of who ate with whom. The sultan ate alone (served and surrounded by pages and sometimes women), the grand vizier ate with the *bash defeterdar* (the finance minister), the two *kadiaskar* (chief judges) ate together, the leftover dishes of the vizier were then served to the chief of the guards, and so on.[24] Women were absent, apart from at the sultan's table. Dinners were silent and rapid. Wine was consumed, often copiously, but not with these formalized meals (except perhaps in the privacy of the sultan's meals) and not publicly and, as such, not part of public conviviality.

Banquets

The occasions for public display of food were the banquets and feasts given by sultans and court notables to celebrate royal weddings and circumcisions. These were occasions of splendid display over several days.[25] Dignitaries, pashas and beys, ambassadors and important persons were entertained at special tables covered with numerous and diverse dishes of opulent food. The janissaries and other forces would be entertained at other locations with simpler but still copious foods. And the general urban populace would be invited to a sugary feast. Suleiman the Magnificent (1494–1566) celebrated the circumcision of his two sons in 1539. Yersimos lists the menu.[26] For the dignitaries, many dishes of pilaf, both of rice and of *rishte* (noodles), yellow with saffron, green with spinach, red with pomegranates, with almonds and raisins, with boiled meats, sweet rice puddings, and so on; then soups, stews, and fricassees, many similar to those on the daily menus but dressed in spectacular forms, with various chicken and mutton stews and pies; copious quantities of boiled and roasted meats and many different birds and game, such as geese, pigeons, and partridges, with pride of place given to peacocks; *bourek* (stuffed pastries) of meat, sweet pastries, fruit (fresh and dried), nuts, and so on. The emphasis is on splendor and display, but ingredients and cooking methods are not radically dissimilar to daily palace meals. The janissaries and soldiery were given great quantities of meat and rice, generously dressed with butter, and much sugar in sweetmeats and pastry (sugar being an expensive commodity, not one of regular indulgence for poorer classes). On occasions, the janissaries would be invited to "attack" and "pillage" the offered foods, to the amusement of the notables. "Pillaging" also featured in the offerings of sugar and other items to the general populace. Elaborate sculptures of spun sugar, portraying animals (familiar and exotic), castles and churches, guns and fortifications, and even human figures (in contravention of religious censor). After a period of display in the hippodrome, outside Topkapi Palace, the people were invited to attack the display, fight and compete to break off and eat pieces of the sweet figures, the spectacle of their melee watched with amusement by the assembled notables. Dishes of rice, meat, honey, and bread were also distributed to the common people.

The French court, from the seventeenth century, and even before, set the tone for fashion and opulence in other European courts and aristocracies, from Italy to England and Russia. French chefs were in demand in all these countries, and famous chefs were a status statement in the most opulent houses, continuing into the nineteenth century and after.[27] This was especially the case in England after the restoration of the Stuart monarchy in 1660. Charles II and James II had spent the years of their exile in the French court and on their return, installed French fashions and personnel, as well as sympathy for the Catholic cause to the English court. This, of course, included food styles and banqueting, and all of this was imitated by the aristocracy.[28] The Stuart monarchy, with its French and Catholic leanings, then fed into the deep schism in English society, following the Civil War and the Commonwealth years of militant Protestantism. This is part of the antagonism to French food that became a recurrent theme in British discourses. They asserted the superiority of plain and wholesome British fare against the sauces and spices—notably garlic—that supposedly disguised the inferior ingredients used by foreigners. These pronouncements are familiar and repeated, for instance, in Robbie Burns's famous "Ode to the Haggis." After addressing the haggis as the "great chieftain of the pudding race," Burns turns to contrast it with dismal French fare:

> Is there that owre his French ragout,
> Or olio that wad staw a sow,
> Or fricassee wad mak her spew
> Wi perfect scunner,
> Looks down wi sneering, scornfu view
> On sic a dinner?

Food styles, then, become ideological issues in the assertion of nationalist and middle-class sensibilities against what is seen as the extravagance and pretension of aristocracies and high bourgeoisie and their French styles. The quest for simplicity and delicacy, however, also had its advocates in France, part of the process of "civilizing the appetite."

We can postulate two different directions of the evolution of ideas and practices of food in France and western Europe from the eighteenth century till the present—the pursuit of elegance in fine dining and the rejection of that in favor of simpler cooking. The first of these was the mutation of the complex courtly cuisine into the haute cuisine prepared by celebrity chefs in prosperous households, grand hotels, and restaurants. Among those were figures such as Carême (1784–1833), the innovator of *grande cuisine*, who cooked for royalty and men of state, Escoffier (1846–1935), who continued and refined the style of his predecessor, then turned to grand hotels and restaurants, and Cézar Ritz (1850–1918) of the famous hotels in Paris and London. They codified grand cooking, devised techniques, and refined the organization of labor in kitchens, practices that were widely imitated and followed; their ideas survive to the present day. Escoffier and others believed that their innovations included s lighter and more delicate "modern" cuisine, in contrast to the heavy sauces and combinations of their predecessors. An important development in the mid-nineteenth century, contributing to greater restraint and delicacy, was the change in the order of the French-style meal to what was known as *service à la Russe*. While *service à la Française* dictated the service of multiple dishes at each course, as we saw, the Russian style (credited to a Russian ambassador to Paris) required the service of individual dishes sequentially, the practice that prevails today. The ready adoption of this method is indicative of the trend to greater simplicity and delicacy. This trend of every generation asserting a *nouvelle cuisine* continued in the twentieth century, when, in the 1960s, there was a movement in French cooking formally called *nouvelle cuisine*.[29] The fine dining of our own time has some continuities with this trend, but with complications to which we shall turn below.

The second trend in the evolution of ideas about food and cooking was the rejection of complex and rich foods, sauces and spices—the doctrine that every ingredient should taste of itself and not be disguised and smothered by those additions. A prominent theorist of this trend was none other than Jean-Jacques Rousseau, as part of his philosophy of glorifying nature. He prescribed fresh, local and seasonal ingredients: vegetables, milk and dairy products, and simple bread. While not vegetarian, he advised little meat. A meaty diet, he believed, was conducive to aggression and cruelty, as demonstrated, for him, by the conduct and sentiments of the English, great meat-eaters.[30] Similar themes occurred in cookbooks and practices of the middle classes in France and in more modest forms in Protestant Holland and England, often in conscious distinction from the elaborate French-inspired cuisines of the upper classes and the fine restaurants.

France, beginning in the seventeenth century, saw the growth of sectors of the bourgeoisie. The centralizing state expanded the bureaucracy and the profession of law. Trade, military-provisioning contracts, tax farming, finance, money lending, and the acquisition of land all added to the ranks of the prosperous middle classes. Some high functionaries and lawyers were ennobled. Some cookbooks, notably that of Nicolas de Bonnefons, *Les delices de la compagne* (1654), were addressed to those classes. It emphasized home cooking and the conservation of the products of farm and garden: fruits and vegetables, preserves, and dairy products.[31] It once again enunciated the principle that ingredients should taste of themselves, stressing purity of taste. While not always followed, this maxim entered the ideology of taste and was often repeated subsequently. The *cuisine bourgeois* became a genre of the French culinary endeavor, related to but distinct from the haute cuisine of the court and the aristocracy. It was this cuisine that emerged into the twentieth century as the mainstay of the distinct and superior culinary culture of France so admired by Anglo-Saxon and other gastronomes around the world.

THE SHIFTING CULTURAL FIELDS OF DISTINCTION

Pierre Bourdieu is the social theorist who has contributed most notably and directly to the field of distinction, especially in a book of that title.[32] The book examines the dimensions of distinctions between classes and groups in matters of taste and lifestyle, including the consumption of material and symbolic products, and food is one such product.[33] Distinction follows differential access to two forms of capital, the economic and the cultural. Bourdieu's arguments are illustrated with the findings of survey research conducted in France of the 1970s. Working classes and rural workers are poor on both fronts, and their tastes and consumption of food and drink follow dictates of necessity and ready appetites. They tend to eat heavy, fatty foods that are also cheap: pasta, potatoes, beans, pork, and pork products. This kind of robust diet is shared, at a higher income level, by the minor business classes, shopkeepers and small employers, who eat more opulently, in price and calories, and include some luxury items such as game and foie gras. The differences between workers and their richer counterparts are summed up by Bourdieu as that between *la bouffe* (everyday grub) and *la grande bouffe* (feasting). This differentiation by economic capital contrasts with the dimension of cultural capital that distinguishes the clerical and professional classes, whose consumption patterns tend, on the one hand, to the ascetic and delicate, with concern for health and slimness, and on the other, to the more aristocratic tastes in fine meats and vegetables. The difference between the clerical and the professional is one of economic capital, but they share in the cultural capital to some extent. There is also a gender dimension. Elaborate and heavy preparations, such as *pot-au-feu, blanquette,* or *daube*—all labor intensive and long-cooking casseroles, typically prepared by wives—tend to be favored by the clerical class and avoided by the professionals in favor of simpler grills, for both labor and health and beauty considerations.

A great deal has happened in Europe and the world since Bourdieu's 1970s France, and while his general framework and emphasis on the dual bases for distinction in economic and cultural capital remain useful, the substantive accounts are dated. Let us look at some patterns of distinction in the modern world. Starting in the 1980s, food and gastronomy emerged as fields of intensified cultural interests and spectacle on a global scale. Not, this time, at the tables of royalty and aristocracy or even in grand hotels, but in the public spaces and spheres of media, travel and tourism, cookbooks and cooking magazines, and then social media, blogs, reviews, and images. French or Italian or Mediterranean dishes were merged with foreign styles and flavorings from wide cultural horizons, such as oriental styles and flavorings. Japanese raw fish dishes, for instance, found not only great popularity at all levels but also many imitators and innovators under European labels, such as *tartare de bar* or *de St Jacques*. Cilantro, ginger, and wasabi cropped up in French classics. Fruit and sweet additions to main dishes, characteristic of the old cuisines of the Middle Ages and the Renaissance that were banished in the quest for purity of taste, have now returned, with mango sauces, pomegranate syrup, sugar, and honey in so many fashionable dishes. The banishing of spices and flavorings in the process of "civilizing the appetite" seems to have been reversed in the globalized spectacle of cuisine. These interests are fanned by books, magazines, television shows featuring celebrity chefs, blogs, social media, and the diversity of restaurants and eateries. At a mass level, we have the emergence and wide popularity of the ready meals found on all supermarket shelves that often feature world foods, from hummus to pizza to Indian, Chinese, and generic "oriental" foods. In such a field, cultural capital resides in specialized knowledges and tastes and in keeping up with trends and fashions, which elevate the "foodie" above the mass of consumers. In their book, *Foodies: Democracy and Distinction in the Gourmet Foodscape*, Josee Johnston and Shyon Baumann study the modern gastronomic "foodscape," exemplified in food writing, mainly in the United States, found in books, gourmet magazines, food and restaurant columns in major newspapers, as well as broadcasts and social media.[34] They argue that the traditional distinction between highbrow and lowbrow is eroded, in that the foodie quest for novelty and diversity includes an assertion of "democratic" interest in the food of diverse cultures and strata, a willingness to "eat anything." Exoticism, the search for ethnic foods and novel tastes, can form part of this down-market attitude. Yet, this democratic assertion is hedged within the cannons of distinction, "quality, rarity, locality, organic, hand-made, creativity and simplicity,"[35] which are accessible only for those with economic and cultural capital. The old snobbery of French haute cuisine may have been eroded, but the new foodie culture brings forth a whole new set of distinctions, still tied to economic capital, but a more complex field of cultural capital also addressed in chapter 12.

CONCLUSION

This chapter has traced different modes of social differentiation and associated patterns of diets and culinary cultures, with a focus on the cuisines of the courts and aristocracies as key elements in the evolution of those cultures and tastes. In Europe, France emerged as the model and exemplar of courtly and aristocratic cuisine, then of haute cuisine. We examined the sociological dynamic driving constant innovation and the search for a newer and more modern *nouvelle cuisine* and contrasted it with the stable opulence of the food of the Ottoman court. In more recent times, this dynamic for constant innovation has been globalized, overtaking French culinary preeminence and establishing food as a medium of spectacle and entertainment.

The chapter followed Elias and Mennel in postulating a long-term process of "civilizing the appetite." The gross indulgence of the Middle Ages—the heaping of the strong flavors of spices, vinegar, sugar, fruit and honey—gradually yielded to greater delicacy and simplicity, and the elegance of fine dining eventually yielded to the doctrine that ingredients should taste of themselves. Is this trend, however, being reversed in our own days, with globalization, the fusion of ingredients and methods, and the rediscovery of spices and strong flavors, including the mixing of fruit and sweet-and-sour tastes with savory dishes?

We have seen how social differentiation in food cultures and the emergence of courtly cuisines were conditional on the expansion of empires and trade, making available to elites the materials, some from distant lands, through which difference and opulence can be marked. Our own time of ever-expanding circuits of trade, agriculture, industry, and, crucially, communication have brought this process to an apex. Innovation and fashion move at a constant pace, and food serves as a medium of difference and distinction of class and status, not only in terms of the obvious differentiation of wealth and poverty, but also, crucially, of cultural capital.

CHAPTER 10

Political Economy

The Global Food System

In recent decades, concepts like food miles (measuring the energy consumed in bringing food from field to fork) or fair trade (allowing consumers to purchase food produced by small-scale farmers according to enhanced ethical and environmental standards) have made the idea of a global food system more tangible. At one level, the definition of a global food system is fairly straightforward; it suggests a structural relationship between the biophysical, socioeconomic and political aspects of planetary life, which, in turn, "implies there is an interconnection beneath the surface of things"[1] as they present themselves in our daily consumption of food. Speaking of a global food system thus draws attention to the totality of underlying social and ecological processes—sowing, fertilizing, feeding, cultivating, harvesting, slaughtering, packaging, transporting, storing, marketing, and retailing—that allow the everyday act of ingesting food and drink.

Yet, for all the contemporary reflection on food provenance, seasonality, ecological footprint, or animal welfare, there is still reluctance in public discourse to think of these as systematically tied to the history and present of agrarian capitalism (or, more colloquially, agribusiness) and the accompanying political dynamics and dilemmas this throws up nationally and globally. As chapter 9 illustrates, politics and ethics tend to be invoked in the context of individual preferences at the point of food consumption, less so as part of collective experiences of food production. This is somewhat puzzling, since modern political economy has, from its inception in the seventeenth century, emphasized that patterns of consumption are deeply intertwined with modes of production. The so-called physiocrats were fixated with volume of arable land determining any state's capacity to feed its population; David Ricardo famously premised his theory of "comparative advantage" in the example of international trade between wine-producing Portugal and cloth-making England; and Karl Marx's notion of primitive accumulation (itself drawn from Adam Smith) was, as we saw in an earlier chapter, concerned with agricultural enclosure and land dispossession at home and abroad as the foundational moment in the creation of a capitalist world market. As a branch of modern social theory, political economy has thus always been concerned with—and in some instances inspired by—the workings of the global food system.

It is the task of this chapter to elaborate on this latter proposition, offering illustrations of how keywords in political economy like *value, free trade, rent,* or the aforementioned *comparative advantage* simultaneously issue from and help to explain the operation of the global food system today. By engaging with concrete examples from the global trade in coffee, we will encounter instances of how states and markets clash and cooperate in reproducing the global food system. We will also be able to give due consideration to the complex interaction between politics and economics (or public authority and private exchange) in this process, highlighting the global regulation of food production, distribution, and consumption across borders, as well as its more recent neoliberal transformation. Here again, contemporary expressions of global or international political economy (IPE), with their theories of "multilateral regimes" or "value chains," can assist in understanding the global food system. To frame these multifaceted discussions, we first turn briefly to the historical evolution of successive global food regimes.[2]

FOOD REGIMES, PAST AND PRESENT

In a seminal 1989 article, Harriet Friedmann and Philip McMichael introduced the concept of the food regime, which, inspired and informed by the French regulationist school of analysis, sought to link "international relations of food production and consumption to forms of accumulation broadly distinguishing periods of capitalist transformation since 1870."[3] Their principal aim in that essay was to chart the interdependence between the industrialization of agriculture and a bourgeoning international system of national states. Friedmann and McMichael posited the transition from a first settler-colonial regime—in which family farms in the temperate "white dominions" of the Americas and Australasia, together with the tropical plantations that produced "drug foods," fed European wage labor by exporting affordable grain, sugar, coffee, tea, cacao, and meat—to a second, postwar food regime driven by U.S. hegemony over an inter-state system that reproduced the American model of agribusiness across the capitalist world.

Whereas the earlier food regime was defined by extensive accumulation within an imperial framework, the latter is characterized by the intensification of agriculture through integration and consolidation within the agribusiness sector and the export of this paradigm across a postcolonial world of formally independent states.[4] During the Age of Empire (1870–1914), a combination of mass settlement in the "neo-Europes," the subjection of colonized lands and people to commodity production, and the subsequent overseas shipment of grain and meat, all extended a previously elitist wheat-livestock diet to working classes in capitalist economies around the globe. After World War II, Washington oversaw a global rule-governed food regime that combined market integration within and between allied states with the protection of domestic food prices through subsidies, quotas, foreign aid, tariffs, and fixed exchange rates, all buttressed by the newly founded Bretton Woods system and the associated U.N. specialized agencies (most notably the Food and Agriculture Organization).

Accompanying these international agreements was a complicated and dilated, but nonetheless irreversible process of corporate concentration in the agribusiness sector, spurred on by the accentuated mechanization of capitalist agriculture and its gradual consolidation into large-scale, intensive, specialized, monocultural farms. With the growing commodification of inputs (sowing and harvesting, pesticides, herbicides, fertilizers, feed, pharmaceuticals, and transport) and the increasing dependence on wholesale of outputs to powerful supermarket chains, vertical integration within the food system yielded the (in)famous hourglass pattern—hundreds of millions of consumers purchase food from a very narrow band of retailers, wholesalers, manufacturers, and processors, who, in turn, buy produce from millions of farmers reliant on literally a handful of input suppliers. Thus, for instance, in the United States, Smithfield Farms has a 31 percent market share of the American pork-packing industry while also raising close to 20 percent of the country's hogs. Upstream, at the supply end of the system, three firms (Cargill, Archer Daniels Midland, and Bunge) control 90 percent of the world's grain; ten companies own two-thirds of the world seed market; and the ten leading firms command almost 90 percent of the world pesticide market.[5] Downstream, at the consumption end of the process, agribusiness has "introduced a manufactured diet whose main components are fats and sweeteners, supplemented with starches, thickeners, proteins, and synthesized flavours."[6] Like other sectors of the capitalist economy, the food system is thus characterized by driving forces of profit, productivity, competition, innovation, product differentiation, and corporate mergers.

In subsequent elaborations on these ideas, Friedmann and McMichael have (separately and as coauthors) refined and modified their schema, introducing a third "corporate food regime" corresponding to the neoliberal era since the late 1970s that witnessed the re-regulation of markets in favor of large agribusiness concerns; the privatization of state-owned or controlled production; a new wave of peasant and small farmer dispossession; an accompanying commodification of biotechnology via genetically modified organisms (GMOs) and seed patenting; as well as the "financialization" of the food system through private and sovereign investment funds and international land transfers.[7] The content and accuracy of this periodization need not detain us here (though, as we have seen in earlier chapters, the settler-colonial food regime long predates 1870, and the corporate regime continues to be more protectionist than the label "neoliberal" allows for). What is of interest for our purposes is how the idea of a global food regime helps to identify specific historical epochs in which "complementary expectations govern the behavior of all social actors, such as farmers, firms and workers . . . as well as government agencies, citizens and consumers."[8] Put differently, reference to a given food regime allows scrutiny of the causal interconnections between the political institutions, market forces, social movements, and environmental determinants involved in reproducing the global food system. On this account, food production and consumption cannot be considered in isolation, nor can trade politics be divorced from agricultural economics, even less so can our diets be disassociated from global climate change. The challenge, then, is to activate our "sociological imagination," as C. Wright Mills once recommended, and disaggregate the empirical research into each of these factors, only to then reassemble them into a systematic analysis of how each historical food regime is reproduced.

FOOD AND POLITICAL ECONOMY

In the long history of political economy, there have been three broad approaches to this endeavor—most of them concerned with food production and consumption, either directly or through associated notions of productivity, rent, value, trade, and competition.[9] The first of these falls under the rubric of "mercantilism." Stretching from the seventeenth-century treatises by Thomas Mun to the more contemporary advocates of "economic statecraft," the underlying assumption in this perspective is that there is a limited amount of wealth in the world, and therefore one political community's gain (be it a nation-state, an empire, or an economic union) necessarily involves its rival's loss. One variant of this thinking, though not strictly speaking mercantilist, was Reverend Thomas Malthus's famous claim that, as population grew exponentially and food supply only arithmetically, famines, droughts, wars, and other such calamities would readjust the balance between people and produce. Much of the contemporary "limits to growth" or "resource scarcity" discourse is inspired by such conceptions of absolute (as opposed to relative) limits to the planet's carrying capacity, with land and food among the various world resources that, according to this perspective, are being irreversibly depleted by population growth.[10]

In some early mercantilist writing, wealth is narrowly defined by bullion reserves, but in its wider acceptance, mercantilist principles simply stipulate that any economy should export more than it imports. Here the political-juridical authority and military-diplomatic power of the state plays a critical role by, among other things, legislating for tariff barriers protecting domestic agriculture or, more recently, mobilizing sovereign wealth funds to lease or buy land abroad in pursuit of national food security. Mercantilist or "economic nationalist" theorists and policies are not flatly opposed to international trade, but they do advocate its regulation in favor of a putative national self-sufficiency.[11] England's seventeenth-century Navigation Acts—obliging all ships landing colonial goods on English shores to be English or owned, built, and manned by an English colony—are a good early example of this, as is the European Union's Common Agricultural Policy. There were significant political implications to both these economic policies, the first contributing to Britain's industrialization through the food surpluses of its overseas empire, the second shoring up electoral support from farming communities across the continent, as well as acting as a cornerstone of European integration.

Mercantilism was the main target (some argue, invention) of the classical or liberal political economy that flourished toward the end of the eighteenth century. Famously expressed in the writings of Adam Smith and David Ricardo, this conception of wealth creation and distribution championed the role of markets in allocating scarce resources most efficiently. One of the great contributions of these authors to political economy was the insistence that it is human labor—and not land or other natural resources—that is the chief source of the market value and, by extension, the price of commodities. The commercial (or capitalist) revolutions Smith, Ricardo, and their contemporaries were witnessing had turned all factors of production (land, labor power, machinery) into alienable commodities, thus permitting their competitive market exchange. Smith's fabled invisible hand, guided by the self-interest and profit-maximizing passion of individuals, delivered the most efficient production and distribution of commodities: "It is not from the benevolence of the butcher, the brewer, or the baker that we expect our dinner, but from their regard to their self-interest."[12] Such commercial interaction extended internationally, with trade acting as an incentive for productivity gains through the specialized division of labor and the accompanying development of comparative advantage between nations. In Ricardo's classic rendition, "Under a system of perfectly free commerce, each country naturally devotes its capital and labor to such employments as are most beneficial to each. . . . It is this principle which determines that wine shall be made in France and Portugal, that corn shall be grown in America and Poland, and that hardware and other goods shall be manufactured in England."[13]

It is important to emphasize how these doctrines of free trade and international division of labor emerged in the context of very concrete political disputes over state intervention in the economy. David Ricardo, in particular, opposed the British Corn Laws, which, from 1815 to 1846, imposed restrictive tariffs on grains imported to the United Kingdom, thereby protecting domestic producers and undermining price cuts. Together with other campaigners, he called for the repeal of the Corn Laws so that free trade and competition could cheapen food, lower wage costs, free up manufacturing capital, and gradually increase employment in productive economic sectors. Here another concept of classical political economy, namely rent, becomes relevant, since for Ricardo and other economic liberals (though in contrast to Smith, as we saw in chapter 5), the part of agricultural surplus diverted to landlords in the form of rent necessarily squeezes the profit share generated through commodity exchange, thereby reducing the amount of capital available for reinvestment and growth. Since then, neoclassical economists have referred to the state-led regulation of markets through tariffs, subsidies, duties, or licenses as potentially "rent-seeking" activities that distort market competition, increase transaction costs, and simply line the pockets of employees in an already bloated public sector.[14] Commodity marketing boards—particularly in postcolonial states—could be seen as the archetypal rent-seeking institutions.

The development of capitalist industrialization during the nineteenth century led to a third broad approach to political economy, namely, Karl Marx's critique. It is often said that Marx's work combined French political theory, German philosophy, and Scottish political economy to deliver a unique explanatory toolkit based around the concepts of class, exploitation, surplus value, and modes of production. For the purposes of this chapter, the most important insight Marx and his followers contribute to our understanding of food, politics, and society is the process of commodification of the global food system. Marx adopted Smith's and Ricardo's conceptual innovations to formulate his own labor theory of value whereby, in the historically distinctive capitalist mode of production, all factors of production (labor, capital, and land) are combined as alienable equivalents through market exchange to create value—generally, though not exclusively, represented in the form of money and profit. Put differently, the capitalist transformation of all things—including the DNA of living matter—into a potential commodity (i.e., an exchangeable value) means that more and more areas of life are subject to the cash nexus. In the case of the global food system, this simply involves the commodification of the whole food chain so as to deliver the historically unprecedented worldwide "price-governed market in an essential means of life" we have encountered at various junctures of this book.[15]

Once again, it is relevant to note that the process of commodification is related, but not equivalent to, the rise of a world market after 1492. As we saw in chapter 3, the long-distance trade in commodities preceded capitalism as a mode of production. It is not the buying and selling of foodstuffs in itself that defines capitalist agribusiness, but rather the production of edible or drinkable commodities through the exploitation of wage-labor working on commercial land, using machinery that is owned or controlled by private companies. Value, on this view, is created only in the process of capitalist production, which requires the commodification of all inputs, or what Marx labeled their "real" as opposed to merely "formal" subsumption to capital. Such distinctions matter because they can help not only to identify the unique dynamics of capitalist agribusiness (competition, productivity, financialization, complex divisions of labor, and so forth) but also, crucially, to contrast these with preceding food systems, such as those of antiquity explored in chapter 2, governed by slavery, tribute, conquest, and munificence, rather than value-creation as such. (Though this does also throw up the thorny issue of whether value is created by contemporary unwaged or coerced labor in households or other units of production). More specifically, this entails seeing the contemporary food system as part of a global value chain in which food and beverages enter the world market as a primary commodity (say, cacao) grown in one country, then to be processed, packaged, marketed, and sold in various other countries as a finished product like a chocolate bar. As analysts of global commodity chains have demonstrated, an international division of labor that distributes the production, processing, and consumption of food across different geographical locations raises the question of where value is being captured or realized in this commodity chain—an issue we will return to presently.[16]

We have seen, then, in this whistle-stop tour of modern political economy, how a host of concepts linked to this tradition of social theory emerged in tandem with the development of capitalism as a distinctive socioeconomic system. Food and drink appear here and there with different degrees of immediacy as elements in the description and explanation of this new "commercial society." But it is really in its concern for the transformation in the relationship between agriculture and industry, land rent and commercial profit, and production and trade that political economy becomes central to our understanding of the modern food system. For all their ideological and methodological differences, the political economists discussed above agreed on one thing: a new type of market society was engulfing ever-larger parts of the world into a logic of profit, growth, competition, and trade that was having a momentous impact on socioeconomic organization and, by extension, the way societies produced, processed, and consumed food. The private enclosure of land, the accompanying emergence of a labor market of landless workers, and the creation of a world economy through conquest, dispossession, and colonization was revolutionizing society's historical relationship to nature as a source of subsistence and nourishment (part of what Marx called the "metabolic rift" between humans and our environment). Classical political economy was thus centrally concerned with what we today understand as "development," the transition from predominantly agrarian societies governed by use-value to industrial civilizations characterized by the mass production of commodities for exchange-value.[17] By the mid-twentieth century, another prominent social theorist of economic history and anthropology, Karl Polanyi, labeled this process the "great transformation," the rise of a "self-regulating market" that nonetheless periodically provoked a political counter-movement in the international, state-led regulation of land, labor, and money. Out of this "double movement" in the late nineteenth century, Polanyi averred, "a network of measures and policies was integrated into powerful institutions designed to check the action of the market relative to labor, land and money."[18] We turn now to a more concrete exploration of how the various concepts associated to this great transformation might help explain the workings of the modern food system.

STATES AND MARKETS IN THE GLOBAL FOOD SYSTEM

One of the principal criticisms of classical political economy and its more recent neoliberal derivations is the idea that markets somehow emerge spontaneously from "the propensity in human nature . . . to truck, barter, and exchange one thing for another."[19] In fact, as we have seen in this and other chapters of the book, the forging of a world market, even more so the commodification of agriculture, was the outcome of war, revolution, slavery, violence, and state intervention. As Polanyi wryly put it, "Laissez-faire was planned."[20] The notion of a self-regulating market, Polanyi once again insisted, is a utopian aspiration that can never be entirely realized because markets (and therefore profits, competition, innovation, and productivity) are always somehow embedded in wider social, political, cultural or religious institutions and practices.

The organization of the global coffee market—since 1945 the world's second most valuable traded commodity—is a good example of this embeddedness. It was originally introduced as a cash crop to the colonial islands of the Dutch East Indies in the early 1700s and then to the West Indies and Americas during the early 1800s, and by the end of that century, 75 percent of the world's coffee was harvested in Brazil. In subsequent decades, other Latin American (principally Colombian) and then African producers entered the market as significant but secondary players compared to Brazil—a distribution that remained intact until Vietnam irrupted into the scene at the turn of the twenty-first century, when it became the world's second largest coffee grower, currently accounting for 20 percent of global output. The everyday reproduction of this lucrative world market is shaped by a number of key international organizations, political institutions, and social agents that together facilitate (or not) the myriad economic transactions that allow the daily consumption of an estimated two billion cups of coffee.[21]

For decades, the most significant of these institutions was the International Coffee Organization (ICO), which, from its foundation in 1963 to its disbandment in 1989, implemented successive multilateral International Coffee Agreements (ICAs) including exporting and importing countries. In essence, the latter aimed at stabilizing (and raising) the global price of coffee through the coordination of production quotas. Member states—by the 1970s, including all major producing and consuming countries—agreed to a set price band and controlled the global supply accordingly. This exercise required an organizational infrastructure that involved managing a budget and permanent personnel; collating and disseminating relevant data; administering a certificate of origins system, thereby enforcing quota adherence; and monitoring export stocks and flows. Far from reflecting a "natural propensity to truck, barter, and exchange," the world coffee market is a political construction. As one analyst with insider experience of the ICAs has suggested, "Within the regulated environment, allocations were not made purely in response to supply and demand and the subsequent formation of prices; rather they resulted from a political process."[22]

IPE scholars have defined such political regulation of markets as international regimes: "Implicit or explicit principles, norms, rules and decision-making procedures around which actors' expectations converge in a given area of international relations."[23] This understanding of *regime* is similar to but slightly different from that of Friedmann and McMichael's *food regime* in that it is more narrowly concerned with the question of why states cooperate in an anarchical international system in which there is no overarching authority to enforce agreements. Different IPE theories offer contrasting interpretations, but, following the earlier tripartite division in approaches to political economy, it would be fair to say that realist or mercantilist approaches tend to emphasize the relative gains that are likely to accrue to the dominant power vis-à-vis its rivals through cooperation. Liberal or institutionalist accounts underline the absolute gains to be made by all participants in international agreements via the reduction of transaction costs and increased predictability of market behavior through trust, iteration, and reciprocity between actors. Meanwhile, Marxist or radical interpretations would drive home the point that international cooperation like that represented by the ICO simply codifies, legitimizes, and, to a degree, deepens an already unequal distribution of power and wealth among states and classes in the production, processing, and consumption of coffee. The advance offered by IPE in relation to many of the dominant neoclassical or neoliberal ideas of economics is the explicit recognition that markets are necessarily embedded in politics—there would, in effect, be no functioning global coffee trade without the political institutions (including the ICO) that regulate and underwrite these market transactions.

The political dimension to markets is further demonstrated in the powerful social and economic forces that accompanied the development of the ICAs. On a national scale, coffee production has historically been organized in very different ways, from slave plantations and smallholder sharecropping to capital-intensive estate cultivation. Various social groups—from Brazilian *fazendeiros* (large landowners) to New York auctioneers, Indonesian smallholders to *Kinh* frontierspeople producers in Vietnam—have thus become major political actors in these contexts. In the course of the twentieth century, but particularly after 1945, producing countries established regulatory bodies—marketing boards and *caisses de stabilisation* in Africa, the system of coffee *institutos* in Latin America—that allowed the state to intervene in different aspects of "upstream" production, from controlling export-import and processing and transportation (in the marketing board model) to the more straightforward provision of inputs and credit (as the *Instituto Brasileiro do Café* did). All these examples suggest that—among producing countries, at least—the coffee industry was too important to be left to the self-regulating market alone.

Yet it was among the main consuming nations, overwhelmingly in the Global North, that another key actor of the global coffee market emerged, namely the coffee-roasting firms. This generic category covers both subsidiaries of global food corporations like Nestlé and coffee-buying and coffee-processing companies like Lavazza, as well as so-called specialty coffee outfits like Starbucks. From the beginning of the global coffee boom in the eighteenth century, commodity exchanges in London, Antwerp, Amsterdam, and later Chicago, Hamburg, and New York contributed to the grading, price setting and trade of coffee. Like other commodity markets, the standardization of the product and its insertion into a futures market was facilitated by the technological innovations in transport, processing, storage, and marketing of food considered in chapter 5. When the coffee-exporting countries clawed back market control during the middle of the twentieth century, roasters in importing nations made sure they leveraged their market power at the consumption end of the chain. Robert H. Bates illustrates how the political economy of the global coffee market operated at ground level when he reports the visits by representatives of General Foods to Bogotá in 1963 to secretly negotiate a reduction in the price of Colombian coffee sold to their firm in exchange for an increase in the country's share of the U.S. market (at that time amounting to almost 25 percent of Colombia's total coffee production).[24] Similar deals were struck with leading American roasters like Procter and Gamble, opening major producers to a greater share of the U.S. market in exchange for price discounts. Such long-term inter-firm agreements buttressed the cross-governmental coffee regime, with American roasters acting as lobbyists for coffee growers in Congress. The upshot, as Bates indicates, was that "the coffee market lay embedded within domestic political institutions and international diplomatic relationships. Within that political context, political service by large roasters constituted economic competition by other means. . . . The economic and the political thus became one."[25]

It is telling that the collapse of the ICA regime after 1989 coincided with the ascendancy of neoliberalism as both socioeconomic project and policy. The geographical diversification of the supply base, free-riding and endemic disagreement over quotas among coffee producers, as well as changing consumer preferences all contributed to the decline of the ICA regime at the end of the past century. But liberalization and deregulation of domestic markets in the course of the 1990s must also be taken into account when explaining its demise. As the state in producing countries withdrew its previous control of the coffee industry (the *Instituto Brasileiro do Café,* for instance, was unceremoniously abolished in 1990), private foreign investment grew upstream, while downstream roasters and retailers increased their hold over the value chain. One consequence of these multiple and variegated transformations was the diversification of producers and final products on the one hand and the concentration of roasters and traders on the other, thereby replicating the hourglass model cited earlier. Recent figures suggest that "the ten largest roasters process almost 40% of all the coffee that is consumed worldwide" while "50% of the world's green [unroasted] coffee beans are traded" by just three companies: Neumann Gruppe from Germany and the Swiss Volcafé and ECOM.[26] Product differentiation through private grading standards devised by specialty coffee firms, technological innovations such as single-serve systems like Nespresso, and the growing popularity of coffee across Asia have, over the past decades, all contributed to a huge worldwide increase in the demand for coffee. Yet these developments have been accompanied by a decline in the farm prices for the primary product, delivering what Daviron and

Ponte have called the "coffee paradox": "farmers are getting a decreasing share of the final price paid by consumers for coffee. This means that the value added (and rent extracted) along the chain takes place increasingly in consuming countries."[27]

The narrow focus on coffee has aimed to offer an illustration of how concepts in political economy can help us navigate through the intricacies of the global food system. Although it is important to underline the particularities of different agri-food chains in the world economy (the coffee industry has not, for example, been subject to the same domination by the supermarket-led "retail revolution," nor has it experienced high degrees of vertical integration present in other sectors), it is nonetheless the case that since the 1995 Agreement on Agriculture (AoA), global agribusiness has been regulated through the multilateral regime of the World Trade Organization. In the closing section of the chapter, we consider ways in which the market-driven but rule-governed globalization of the food system has been accompanied by new phenomena such as land grabs, food financialization, fair trade, and campaigns for food sovereignty. It is suggested that all of these can be explained with reference to concepts generated by the various traditions of modern political economy.

A NEOLIBERAL FOOD SYSTEM

At root, neoliberalism combines a historical era, an economic doctrine, and a political project. Pioneered under Pinochet's military dictatorship in 1970s Chile, the principle of the "free economy and the strong state" was subsequently implemented in Organisation for Economic Co-operation and Development (OECD) countries throughout the 1980s and 1990s with successive electoral mandates and, in the Global South, through International Monetary Fund and World Bank sponsored Structural Adjustment Programs. Cold War politics, which had made the West fearful of peasant-based developing-world revolutions benefiting the Soviet bloc, deliberately excluded agriculture from the postwar General Agreement on Tariffs and Trade (GATT), and therefore largely protected this sector from the early stages of the neoliberal counter-revolution during the 1980s. The collapse of communism—combined with the rise of a liberalizing coalition of small- and middle-sized agro-exporting countries known as the Cairns Group,[28] allied with the concerted lobbying of increasingly powerful multinational agribusiness companies—eventually led to the incorporation of agricultural commodities into the newly founded World Trade Organization (WTO), thereby "set[ting] in place, for the first time, multilateral rules restricting the sovereignty of government to establish their own agricultural policies."[29]

The paradox here, of course, is that state authority invested in multilateral organizations was the principal tool used to restrict state sovereignty over agriculture and integrate it into the global market. Neoliberal globalization, in this sense, was authored by sovereign states. The 1995 AoA sought to increase foreign access to domestic agriculture through the process of "tariffication"—reducing non-tariff barriers such as quotas and replacing controls over *volume* of imports (quotas) with a standardized tariff measurement by *price* at border, thus indirectly lowering tariffs. It also required the reduction of domestic supports and export subsidies, all aimed at facilitating free trade and comparative advantage in ways that Smith and Ricardo might themselves have recognized.

This is not to say that the process was always and everywhere successful—quite the opposite. As numerous studies have documented, the AoA was riddled with caveats and exemptions (mainly favoring the United States and the European Union) to the extent that it ended up reinforcing the very imbalances in the global food system and the inequalities in world agricultural-trade rules it had set out to address in 1995.[30] One result of this entrenchment in the double standard of liberalization for some but protectionism for others was the collapse, in December 2015, of the fourteen-year-old Doha Round of world trade negotiations and the accompanying rise of regional free trade agreements across the north Atlantic and the Pacific rim and, indeed, the bi- and multilateral South-South agreements between developing nations.

The past three decades, then, have witnessed the culmination of a revolutionary process of agrarian change that began in the seventeenth century, what classical political economists understood as the "improvement" of land and the staged progress toward civil or commercial societies and, more recently, has been studied as "development." The so-called agrarian question (how does capitalism transform the countryside?) has found response in the commodification, industrialization, and therefore mechanization of agriculture to an extent that makes it ever harder to distinguish between town and country, urban and rural, and proletarian and peasant in most parts of the world today. We have entered, during this century, what Philip McMichael has labeled a third, "corporate" food regime, characterized "by the global de-regulation of financial relations, calibrating monetary value by credit (rather than labor) relations—as practiced through the privatizing disciplines internalized by indebted states, the corporatization of agriculture and agro-exports, and a world-scale casualization of labor."[31]

This has, to be sure, been a socioeconomically multifaceted, politically contested, geographically uneven, and historically protracted process, without any predetermined destination. But it has equally been a transformation with a specific direction, guided by the generally convergent interests of powerful states and social classes in deepening and widening the commodification of everything and the production of "food from nowhere." The term *food regime* is useful in this regard because it helps by joining up the different parts of the food system, from production to consumption, as well as by "refocusing from the food commodity as *object,* to the commodity as *relation,* with definite geo-political, social, cultural, ecological and nutritional relations at significant historical moments."[32] Such emphasis on the social relations underpinning the global food system, moreover, bolsters an axiom of modern social theory, namely that humans are both agents and products of our immediate environment. What this means specifically for the global food system can be fleshed out, by way of conclusion, with reference to three contemporary aspects of the corporate food regime.

The first of these is the revolution in food retailing manifest in the rise and consolidation of multinational supermarket chains. Economies of scale (in volume processed) and scope (through product differentiation), reinforced by geographical expansion and company mergers, have allowed leading food retailers like Walmart, Carrefour, or Tesco to wield enormous power over their global supply chains. By concentrating access to the bulk of consumers and by controlling different stages of the chain—from sourcing to "slotting fees" for prime shelf location— a handful of supermarket chains have become critical to the pricing, quality, and selection of household groceries across the world's largest economies. At the point of production, this retailing revolution has extended and intensified the capitalist mechanization, commodification, and proletarianization of agriculture, but it has also introduced a just-in-time, "permanent global summertime" supply chain highly dependent on a seasonal, part-time, nonunionized, and generally low-paid industrial reserve army of labor, which also extends downstream to transport and catering. At the point of consumption, supermarket customers have benefited from an explosion in product range at cheap prices. Yet this has been premised on the "obesogenic" environments and carbon-emitting diets associated with the sedentary; car-dependent; meat-oriented; and high fat-, sugar- and salt-content eating habits that contribute significantly to the development of type 2 diabetes, hypertension, and coronary heart disease discussed further in chapter 11. As these "diseases of affluence" extend to the urbanizing developing world, the figure of the recently displaced diabetic fruit picker from the Global South or the low-wage immigrant kitchen porter with heart disease is likely to become commonplace in the neoliberal food system. This shift from an imperial and then Fordist food regime—among whose pathologies were colonial famines or absolute scarcity, respectively—to a post-Fordist corporate food regime where superabundance and waste are chief sources of political and socioeconomic malaise, all reflect the critical importance of identifying the "interconnection beneath the surface of things" when studying the global food system.

The various food justice and food sovereignty movements that emerged in the 1990s constitute a second type of response to the corporate food regime. They signaled the flip side of a Polanyian double movement in their challenge to the neoliberal marketization of land and labor and in their efforts to re-embed these values into the sociopolitical structures of the state, workplace, or community. The terms *food justice* and *food sovereignty* are, of course, shorthand for a multiplicity of social movements and organizations, with a diverse (and sometimes conflicting) set of aims, resources, tactics, and ideological standpoints. But they do sit along an identifiable spectrum of activists, advocates, and concerned citizens who seek variously to radically transform, merely reform, or simply offer ethical alternatives to the existing food system.[33]

Fair trade emerged in the late 1980s as a scheme directly connecting producers and consumers through an alternative commodity supply chain governed by higher ethical, ecological, and quality standards. Its objective involves "cutting down the number of middle agents in the international exchange of commodities," thus enabling "consumers in the North to participate directly in the rebalancing of the world trade system by allowing them to choose to pay more in order to support the livelihoods of farmers in the South."[34] There is no attempt here to socialize or de-commodify agribusiness, but rather to steer commodity markets toward a more economically stable, socially equitable, and environmentally sustainable trade system through the mobilization of consumer sovereignty. Fair trade uses many of the mainstream marketing techniques, such as certification schemes and product differentiation, to increase both ethical sales and consumer consciousness, and it has been successful in occupying supermarket shelf space with some specialty food commodities, notably coffee, tea, chocolate, and bananas.

Proponents of food sovereignty, on the other hand, make a more explicit and frontal attack on the neoliberal corporate food regime by prioritizing "local and national economies and markets" and ensuring "that the rights to use and manage lands, territories, waters, seeds, livestock and biodiversity are in the hands of those of us who produce food."[35] Championed by the international peasant's movement *La Via Campesina,* the idea of food sovereignty crystallized in the mid-1990s as part of the "alterglobalization" movement that was mobilized against global neoliberalism at the successive world and regional social forums—most famously during the 1999 WTO summit in Seattle.[36] In more recent years, advocates of food sovereignty have won elected office, particularly in "Bolivarian" Latin America, where left-populist governments have launched policies and initiatives that aim to implement, nationally and regionally, a program that challenges the existing global food regime.

Lastly, straddling the more ethical fair trade and the more political food sovereignty approaches are campaigns for world trade reform. The emphasis of initiatives like Oxfam's 2011 GROW campaign is on changing the rules governing the global food system so that financial speculation can be reined in, terms of trade rebalanced, and climate changed averted.[37]

The success of these various counter-movements has been mixed, to say the least, but one largely unintended consequence of this activism has been the mainstreaming of fair trade and the invocation of climate change by powerful sovereign and private equity funds in their quest for food security and sustainable development. The latest reaction to the corporate food regime could therefore be seen as the corporate food regime eating itself up—albeit through the greening of capitalism. Harriet Friedmann identifies this new "environmental-corporate regime" with the selective appropriation of "demands by environmental movements for reduced pollution and depletion," and the accompanying reframing of different parts of the global food system with labels like "organic," "bio," "natural," "health," "low-fat," "no added sugar," "responsible," "fresh," "local," or "sustainable."[38] Friedmann's main point is that green capitalism can reconcile profitability with some aspects of the food justice and food sovereignty agendas and, in so doing, re-naturalize the commodification of the food system. Once a given regime consolidates, Friedmann argues, it need no longer be named—as the principle of sustainability is normalized, the green credentials of major food companies are taken for granted.[39] The farm assurance certification system Global G.A.P. (good agricultural practice), initially launched in Europe in 1997, is a leading example of this turn toward the public-private, or "hybrid," governance of the food system according to superior socio-environmental standards. It represents one way in which the combination of consumer activism, increased risks in an ever more complex food system, and the concentration of retail market share can quite rapidly generate new standards and audit mechanisms attuned to environmental, social, and animal-welfare concerns.

While large supermarket chains have been at the forefront of greening capitalism at the retail end of the market, diverse public bodies, private capital funds, and commodity traders have mobilized climate change to justify radical transformations on the supply side of the system. So-called biofuels, in particular, have expanded the frontiers of global agribusiness in the name of reducing carbon emissions and simultaneously opening new horizons for financial investment in arable land and agricultural commodities. Over the past decade, (inter)governmental emissions targets, combined with carbon trading, oil and food price hikes, and the 2008 financial crisis have witnessed the spectacular growth of transnational land transfers (also known as "land grabs") aimed variously at securing food supplies for food-importing countries; generating returns for private pension, hedge, or sovereign investment funds; and/or cultivating ethanol- or biodiesel-producing crops like corn, sugarcane, soy, or *jatropha*.[40] Thus, the complex (and deeply unequal) interdependence of the global food system becomes fully apparent in the feedback loops between climate change, agribusiness, natural resources, (inter-)state authority, and new financial instruments.

This chapter has argued that the wider tradition of political economy offers a useful set of categories for untangling and explaining the complexity of the modern food system. The study of capitalist development has been placed at the center of this analysis, as has the notion of the global food regime as a historical moment of hegemonic stability in the system. Yet as the closing paragraphs of the chapter suggest, any given food regime is always subject to contestations, contradictions, and therefore changes and adaptations. Today's global food system continues to be a structured but dynamic and open-ended process, driven by collective and individual power relations and public and private forces that can result in a variety of outcomes. It this tension in our planetary food predicament between structure and agency—fate and voluntarism—that has exercised modern social theory from its origins.

CHAPTER II

The Self

Food Choices and Public Health

Most books on food and culture at some point feel obliged to deploy Jean Anthelme Brillat-Savarin's over-quoted aphorism, "Tell me what you eat: I will tell you what you are."[1] We have saved it for this eleventh chapter as it speaks so directly to the notion that in many more senses than the purely biological transformation of nourishment into energy, we are what we eat. As was briefly discussed in chapter 6, we betray our conception of our selves and what we project onto others in our public and private eating habits, in the bodies that ostensibly result from those habits, and in our food-related interactions and negotiations—both narrowly intimate and more openly collective—that we carry on every day. We also saw in chapter 7, in relation to alcohol, how most of us live in cultures with loudly articulated government health, nutrition, and food policies, receiving a barrage of messages from public bodies and media sources about what we as responsible citizens ought to eat— and what, in that context, we ought to be. We are told that we are living in a time of global crisis as far as our dietary intake and body size is concerned, and we are surrounded by conflicting reasoning and advice from different quarters, all of which seek to influence our food choices and "would tell [us] what to eat, despite the absence of evidence that dietary advice has done any good."[2]

Rachel Laudan points out that the "culinary determinism" of "you are what you eat," which now generally refers to relatively comfortable ideas about food and identity, was once meant literally and came with moral judgment. Along with many other contemporary scholars, she suggests that it continues to do so in "discussions of obesity, childhood malnutrition, and the like."[3] In this argument, culinary determinism combines with power relations to make a culinary politics. Building on this and, to some extent, on the moral ideas we touched on in chapter 4, this chapter will therefore discuss food and the self through the lens of what we are told by many sources is one of the world's most pressing crises: obesity. We will consider how the current prevailing framing of obesity as a disease developed in the later twentieth century, how a "political economy" of obesity has emerged,[4] the power imbalances inherent in the arguments over diet and body size, and the alternative frames for considering obesity and fatness that have been proposed.[5] The development of the individual's relationship with food as a matter of public interest and as a subject of government advice and national health policies, as well as the reality of personal choice in these contexts, will be explored through a brief review of national dietary guidelines from various parts of the globe. All this points to the emphasis we place throughout the book on eminently modern structures and processes—states, markets, individual autonomy, consumer sovereignty, globalization, and commodification—in conditioning patterns of food production, preparation, and consumption. Here modern social theory and its attendant concepts of anomie, policy, or the psyche prove once more indispensable in explaining and understanding our food choices and their relation to public health.

THE SELF AND SELF-CONSCIOUSNESS

Before the modern era, the individual was most generally conceived of primarily as a collective being: part of a community or a member of a group. One of the earliest philosophers to consider the individual in terms of a "self in oneself" was René Descartes (1596–1650), whose conception of the individual as a "thinking thing" was emphasized through separation between the mental and the physical in the world in general, reflecting an inner separation of mind and body. John Locke's *Essay Concerning Human Understanding* (1689) specifically conceptualizes consciousness in the modern sense, developing and challenging Descartes's concept of the self. According to Locke, instead of being founded on either the soul or the body, personal identity "depends on consciousness, not substance." As Etienne Balibar puts it, the self in the Lockian sense depends on a notion of self-consciousness that includes "the functions of intellectual vigilance as well as those of responsibility and of 'property in oneself.'"[6] There is a distinction between personal identity and individual identity, and self-consciousness is active engagement with one's own thoughts, responsibilities, and place in the world.

The self is increasingly studied as a neuroscientific question, and in this chapter we work with longstanding notions of the self as a philosophical and, since the late nineteenth and twentieth centuries, a psychological concept. We live in a world where "we are who we feel ourselves to be."[7] In other words, we construct ourselves and create the character we appear to be or wish to appear to be, and, equally, we are influenced in our perception of what that self is by the projections of the outside world. Expanding beyond Descartes's idea that we think therefore we are, contemporary thinkers like Balibar suggest that "being a self means that in addition to having an answer to the question 'Who am I?' you can answer it in a particular way."[8] This is an additional degree of self-consciousness than that of personhood, and it implies a constant self-vigilance and self-monitoring, through which we relate to the present and the past. The examples considered here therefore ask the reader to think about the self as integral to cognition, affect, personal motivation, and social identity. When it comes to food, we consider, in particular, how our conceptions of the self impact upon personal and social identity and purpose with regard to what we consume and how we feel and think about our bodies; how this influences our approach to our environment and to other people; and, in this context in particular, how the self relates to the other.

Psychologists argue that it is in our capacity for "thinking otherwise" that we can most usefully confront ourselves. In the process of achieving a definition of our selves and interpreting that self, we consider our own unique characteristics, our closest relationships, and the groups that we belong to or aspire to belong to. To differing degrees, in each form of self-realization, the "individual self is achieved by differentiating from others."[9] We are constantly comparing ourselves with others in our close and remote circles, and it is this internal reflection on outside influences that shapes our responses. Sociologist Alan Warde argues that our approach to food is one of the classic areas in which this process most frequently shifts individuals "towards self-discipline and self-surveillance. It is not a matter of external compulsion, by peers or the state, that persuades one to eat carefully, but an internalized and self-regarding regimenting of consumption."[10]

To begin, we return to Brillat-Savarin's *Physiologie du Goût (The Physiology of Taste)*, considered by many to be the pinnacle of popular nineteenth-century French gastronomic writing. First published in 1826,[11] it has remained in print ever since, in multiple languages. Brillat-Savarin situated himself firmly in the bourgeois present, proclaiming that he was writing about contemporary gastronomy as "the intelligent knowledge of whatever concerns man's nourishment."[12] His stated mission was to consolidate all known wisdom associated with food and eating, allying gastronomy in a modernist vein with natural history, physics, chemistry, cookery, business, and politics. In doing so, he intended to clearly distinguish gastronomy from gluttony and greed and to elevate it into a science, beyond the straightforward acquisition of knowledge or the development of a certain kind of expertise.[13] His highly readable, wittily avuncular style allowed him to do this by directly appealing to his bourgeois readers' sense of self (including their snobbery, which we'll return to in a brief discussion of his "gastronomical tests" in chapter 12, "Consumption").[14] He takes the idea of gastronomy as branch of science into a quasi-learned sphere where, besides developing a (not entirely convincing) theoretical chemistry of food and its essential components, he meditates on the senses of taste and smell and discusses in detail the personal and social impacts of food and eating, including, notably, obesity and thinness.[15]

For Brillat-Savarin and his nineteenth-century audience, extremes of body size, whether excessively thin or excessively fat, were seen as afflictions that might either be caused by illness—especially in the case of extreme thinness—or by poor personal habits. In both cases, the body on view was taken as a clear reflection of the self within. Reflecting a sentiment that still lies, problematically, beneath much of the moralizing tone endemic to contemporary discussions of obesity as a personal failing contributing to a social crisis, he bluntly asserts that "if obesity is not a disease, it is at least an unfortunate indisposition, into which we nearly always fall through our own fault."[16] It is evident that derogatory language of blame, a form of fat-shaming, is not a recent phenomenon.[17] Since, Brillat-Savarin claims, "obesity is never found either among the savages, or in those classes in society in which men work to eat, and eat only to live,"[18] it is our obligation to "cure" obesity should we fall prey to it, through a combination of "discretion in eating, moderation in sleeping, and exercise on foot or on horseback."[19] His definition of obesity seems to be one of real extremity—referring to mythical Roman emperors so fat they wore their wives' necklaces as rings—but Brillat-Savarin's ideas about plumpness, fatness, and obesity fall into a pattern that remains all too recognizable in many parts of the world today, particularly the Global North.[20] We should pause to remind ourselves that his comments were published almost exactly two hundred years ago. How is it that current public discourses of disease, flawed choices, and personal responsibility in the face of so-called public crises of obesity and eating disorders retain so much in common with the musings of this middle-class nineteenth-century gentleman?

OBESITY AND LANGUAGE

The use of language is one of the potentially problematic components in discussions of body size, eating habits, and conceptions of the self. Many scholars in the field of fat studies indicate their rejection of the terms in which recent debates have been framed by adopting the use of scare quotes when using *obesity, overweight,* and other associated terms. This chapter does not do so, but this is an aesthetic choice made to save the reader from a text bristling with quotation marks and should not be read as unquestioning acceptance of a prevailing public health discourse (or of Brillat-Savarin's social commentary). This discussion assumes that readers understand that "all forms of dietary discourse [are] social constructions"[21] and sets out to explore what that might mean within past and present public discourses that have sought and continue to seek to shape our food choices.

According to the *Oxford English Dictionary,* the noun *obesity* has been in use in English since at least 1611, taken from the classical Latin word *(obesitas)* via the middle French word *obésité* (1550). It applies to "the condition of being extremely fat or overweight; stoutness, corpulence."[22] It seems safe to assume that people have always noticed variations in one another's body size and used language to describe it; we know that those at the extremes, whether fat or thin, have been tropes of art and fiction for centuries. Long before Jack Sprat (who, according to the nursery rhyme, could eat no fat) and his wife (who could eat no lean), we have countless images portraying the wealthy fat and the skinny poor, a classic example being Pieter van der Heyden's 1563 engravings, after Breughel, *The Fat Kitchen* and *The Thin Kitchen.*

Predictably, the former is bursting at the seams with plump, well-dressed and well-fed people at a table groaning with sausages and breads, surrounded by bubbling pots, meats roasting on spits before a generous fire, with an ample mother breastfeeding her chubby child in the foreground, while in the background a thin, ragged man is being kicked out the door. *The Thin Kitchen,* on the other hand, shows a scrawny collection of ragged people scrambling for a few scraps from a basket or dish on an otherwise empty table, one tiny pot over the meager fire, a child

FIGURE 8. *Fat Kitchen,* by Pieter van der Heyden, 1563. *"The Fat Kitchen,"* 1563. *Engraving by Pieter van der Heyden after Pieter Breughel the Elder. Courtesy of the Metropolitan Museum of Art (under creative commons). (See https://metmuseum.org/art/collection/search/392424).*

upending an empty saucepan over its head, and a bony, slack-breasted old woman feeding an equally emaciated infant via a feeding horn, while two of the company desperately try to prevent the portly rent collector from coming through the door. Body size and its associated foods (and even cooking methods) are utilized in these images in ways that were consistently recognizable across four centuries as being characteristic representations of rich and poor.

In Breughel's time, plumpness was most often associated with wealth and thinness with poverty. As now, both were tied up in firmly embedded moral ideas. In the Christian model, the sin of gluttony, discussed in chapter 4, was a deadly but not a mortal sin; however, greed, or the appearance of greed, was a transgression to be guarded against. To retain the moral high ground, the well-fed bodies of the wealthy needed, as in Brillat-Savarin's early nineteenth-century model, to retain an acceptable, healthy plumpness rather than teetering over the brink into excessive

FIGURE 9. *Thin Kitchen*, by Pieter van der Heyden, 1563. *"The Thin Kitchen," 1563. Engraving by Pieter van der Heyden after Pieter Breughel the Elder. Courtesy of the Metropolitan Museum of Art (under creative commons). (See: https://www.metmuseum.org/art/collection/search/392426)*

fatness. A glance at, for example, the satirical cartoons of James Gilray in late eighteenth-century England make it clear that judgments on a person's character, such as uncaring, wasteful extravagance (in the enormous, gouty, gourmandizing person of the Prince Regent) or stingy public abstemiousness (in the skinny, parsimonious Prime Minister Pitt) are readily conveyed instantly through their depicted body size. In current obesity discourses, class, wealth, and body size associations have largely reversed—the poor are more likely to be fat, and the wealthy thin—but the moralizing nature of the arguments and language persist.

The unfavorable and logically baseless assumptions we frequently make about others based on appearance are uncomfortable to confront, but countless studies have borne out (and numerous business-training courses repeated) the 1960s research findings of Albert Mehrabian, professor emeritus of psychology at UCLA, demonstrating the extent to which appearance shapes our views. How we look, move, and speak matters far more than what we actually say; in deciding whether we like someone, 55 percent of our view is built on appearance, while 38 percent is related to tone of voice, and a mere 7 percent to the actual words.[23] More recently, the findings on the subjects of body language and appearance of social psychologist Amy Cuddy, who teaches one of the world's most competitive MBA courses at Harvard Business School, have also been widely promulgated, despite the fact that some of Cuddy's former research collaborators have withdrawn support from aspects of her conclusions. Building on the idea that nonverbal communication is one of our most powerful social cues, she suggested that everyone (and women in particular) can be perceived and perceive themselves to have more presence and impact simply by changing their appearance through body language.[24] It is not a great leap from body language to body shape and size, as shown in innumerable other studies demonstrating discrimination in social spaces in general and specifically in the workplace against people considered to be less attractive, especially those with fat bodies.[25] The injustices that arise from this discrimination extend from the workplace to health provision and from interpersonal relationships to economic opportunity to an extent that requires strengthened legal protections.[26] Our bodies and how we use them, the impact our own perceptions of them have on our behavior, and how we are perceived by others as being defined by them are prime components in the determination of the self and one's life chances.

The moral, behavioral, and personal choice–related aspects of obesity are clearly still present in contemporary discourse, supported by the framing of it as a public health crisis in which obesity is a medical issue and body size attributed to the manifestation of problems associated with calorie-dense diets and sedentary lifestyles. Indeed, since the 1980s, the medical definition of an obese adult is one with a body mass index (BMI) of 30 or above, while those between 25 and 30 are considered overweight.[27] BMI as a measure of obesity in adults has been challenged, but it is still fairly consistently used (other methods, such as skin-fold measurements, proximity to an ideal weight, waist-to-hip ratio, waist circumference, and weight in relation to a reference population were used previously). In children and young people, BMI alone is rarely applied. Particularly for younger children, comparisons are made to an "ideal weight" chart; in others, skin-fold measurements are taken. The World Health Organization (WHO) measure determines that a child is obese if he or she is at more than two standard deviations BMI for age and gender within a comparator population, and overweight if at more than one standard deviation. Various regional health organizations have developed calculators for determining whether or not local populations are obese, with more refined models usually developed for children and teenagers than for adults.[28] But what does this definition actually mean or achieve?

To work through a hypothetical example, the U.S. Center for Disease Control and Prevention's (CDC) calculator for children and teens calculates that a ten-year-old boy of 3 feet, 4 inches (100 cm) height and weighing 55 pounds (25 kg) has a BMI of 25. In an adult, that would make him overweight (although the same numbers used in a model developed by an Indian obesity charity would congratulate the same adult on being in the normal range).[29] Our imaginary American child, however, is placed into the 97th percentile of other children of the same age and gender, defining him as obese.[30] All such measures are, of course, subject to "the myth of the average," and to further complicate the idea of such definitions, BMI has been criticized as having been devised by and for Caucasians, meaning that, at best, it applies to a particular body type and is not suitable for global application or as a basis for global comparisons. A chart for Asians and Asian Americans places the cutoff for obesity at a BMI of 27, and overweight at 23, lower than that defined by the National Institutes of Health (NIH) for the general population.[31] These cultural and race-based projections onto ostensibly neutral scientific data are troubling aspect of the medicalization of obesity.

Most countries now calculate the medical costs related to obesity (approximately 7–8 percent of U.S. health expenditure and 4–5 percent in France), but the definition of obesity as a medical condition in itself is problematic. To arrive at these costs, obesity is framed as the likely cause of other serious health problems, such as type 2 diabetes, hypertension, cardiovascular disease (including liability to stroke), and musculoskeletal problems. It is also linked to the incidence of some cancers, as well as having psychological impacts, such as depression and low self-esteem, particularly in adolescents. However, the recommended "treatment" for the condition of obesity itself is much more of a lifestyle intervention than a typically medical one. For example, the CDC advises for children that they should practice healthy eating and drink water, participate in physical activities every day, sleep "enough," and limit the amount of television watched. "Treatments" in adults may be more extreme and include medical interventions such as the various bariatric surgical procedures that limit food intake or reduce absorption, but the basic lifestyle advice also applies to them. More recently,

196 | Chapter Eleven

work on the gut microbiome has led to surgeries designed to reintro-
duce lost microflora that have positive effects on severe obesity.[32] In all
cases, it is the diseases one is apparently more prone to as a result of
obesity that account for the greatest proportion of healthcare spending.

In reports by national bodies, emphasis is also given to the wider
economic costs of obesity. Governments are concerned not only about
the present and future medical costs associated with the conditions
already mentioned—projected to reach £10 billion per year for the
United Kingdom's National Health Service (NHS) by 2050—but also
the wider costs to society of the increasing numbers of obese people
who are, for example, unfit to work and thus eligible for disability payments, an additional cost estimated in the United Kingdom to reach
£49.9 billion per year in the same timeframe.[33] These factors combined
have given incentives to political leaders to focus on the prevention and
management of obesity and its consequences, not only for individuals
but also the wider population. It has also driven them to seek explanations for the "epidemic" and its causes.[34]

OBESOGENIC ENVIRONMENTS

In the early 1980s a new adjective came into use to elaborate on the
causes and effects of obesity: *obesogenic*. It is used to describe social,
environmental, and cultural factors that are thought to tend to cause the
"illness" of obesity, such as inadequate exercise or lack of access to exercise facilities; the suburbanization of societies and reliance on automo-
biles; the wide availability and advertising of high-calorie foods; lack of
access—through poverty or poor quality of local shopping facilities—to
a wide enough range of "healthy" foods; low interest in or understanding of food and nutrition; lack of consumer understanding of positive
food choices due to poor or misleading food labeling; the globalization
of food cultures leading to an incomprehensible "dietary cacophony"[35];
the demise of "home cooking" and "the family meal" (this factor especially found in lamentations in the conservative press); and even a food-
focused version of Émile Durkheim's *anomie* (the modernism-induced
breakdown in the relationship between individuals and society), memorably characterized by Claude Fischler as "gastro-anomie."[36] It is obvious that many of these so-called obesogenic factors particularly apply in
situations of poverty, a not-quite-explicit reversal of historical norms of
the emaciated pauper in a culture in which "contented obesity used to be
a sign of affluence and physical well-being"[37] to one in which, we are

told, "obesity has become the visual manifestation of unhealthy lower-income eating."[38]

The link between poverty or low income and poor nutrition, including obesity, is explicitly made through the concept of "food deserts." This term was coined in 1995 by the working group of the U.K. Low Income Project Team, which investigated the access of disadvantaged groups to sources of healthy food.[39] Numerous reports have been commissioned since, and the concept has been adopted by other public bodies, such as the USDA, which defines food deserts through measures of access—the number of food stores and the individual's distance from them; the individual's access to transport or sufficient income to be able to use them; and the average resource levels for the area. These data are published on an interactive map through the USDA economic research service.[40] A U.K. mapping exercise was undertaken by Dr. Hillary Shaw in 2003 and has been kept up to date until at least 2015 (at time of writing),[41] It uses measures such as the question "do local shops sell 10 or more items of fresh produce?" Shaw found that around 20 percent of rural areas and 25 percent of urban areas were food deserts, where people had to walk more than five hundred meters to reach an adequate shop. According to Lang and Carraher, these food deserts generally arise as a result of planning failure, market failure, and a consumerist approach to health policy that automatically disadvantages those with lower incomes.[42] People with limited access to, in particular, fresh produce are made more dependent on processed, packaged food.

Food technology is a key component in this so-called obesogenic environment and is seen to be to blame in a variety of ways. Processed foods with long shelf lives have become increasingly available, and their sales have been steadily increasing in the last few decades. The "time-cost" of food—the time it takes to acquire food that is ready to consume—has thus decreased, leading to something referred to as "hyperbolic discounting,"[43] in which the convenience of access to prepared foods allows people to eat more of them and more often, whether as part of a meal or in the form of snacks. Studies have shown that among the demographic for whom time spent on food preparation has decreased most dramatically, weights have increased most considerably.[44] In other words, the suggestion is that excessive dependence on conveniece foods can make you fatter. This effect is exacerbated by the content of the food. The practices of adding fat, sugar, and salt to food to make them more appealing to consumers has been widespread and effective. In a 2003 study, researchers found that countries with the most stringent and greatest number of food- and agriculture-related laws and regulations had the lowest incidence of obesity.[45] While labeling has required manufacturers to inform consumers about the ingredients of their products, it is often ignored, misunderstood, or simply unclear to most of them. The decision in the United Kingdom to impose from April 2018 a "sugar tax" on soft-drinks manufacturers notwithstanding, protective practices often also promote industry's interests over consumer interests, or have unintended consequences. For example, the U.S. dairy and corn industries have long been criticized for influencing government policy against consumers' health interests,[46] while food manufacturers' vast advertising budgets mean that consumers, including children, are constantly exposed to irresistible images of food that may not be the healthiest options. As Julie Guthman argues, "Those who want to redress the problem put a great deal of effort into educating people to making better choices rather than into reforming the policies that allow bad food to be produced or mitigating the consequences for those most harmed," but our choices are made in a wider context over which individuals actually have very little control.[47]

Alongside the growing evidence of issues within the food system itself, the much stronger discourse of individual responsibility prevails. Closely allied to the moral arguments of our forebears, this is a discourse of moderation and the exercise of good choices—we are not obliged to purchase unhealthy, fattening foods just because they are available, so their ubiquity is suggested to be an issue only for individuals lacking self-control. Nonetheless, researchers have found clear similarities between overeating processed foods and addiction to smoking and alcohol, and this ties into difficulties of public health messages allied to behaviors without immediate negative consequences.[48] People may be aware of possible negative impacts and may even intend to do something about the situation—like impose the "cognitive restriction" of a diet[49]—but this can endlessly be deferred, especially in the face of easy access to the desired object. Furthermore, research also shows that we do not gauge our food intake consistently or accurately, especially when snacking. For example, when driving or watching television, our self-monitoring of food intake decreases.[50] Across the Global North, levels of physical activity have reduced considerably, whether due to reliance on automated transport, the reduction in school sports, or the fact that fewer people do physically demanding work. In many countries, the car has become the primary means of transport.[51] We increasingly are more sedentary both in our work and leisure activities.

Obesity is no longer considered to be primarily an affliction of the wealthy and of wealthier nations, but rather a product of environmental factors that apply to the global food system.[52] According to the WHO, overweight and obesity have risen and are continuing to rise dramatically in low- and middle-income countries, especially in urban settings.[53] Children are seen to be particularly at risk in these countries, where, in absolute numbers, more of them are overweight and obese. In 2014, 48 percent of overweight children under the age of five lived in Asia and 25 percent in Africa.[54] In adults, too, there have been dramatic changes in which groups are perceived to have the greater problem. Once a condition of affluence, particularly associated with wealthy males in developed societies, obesity is now associated with poverty in both genders, but especially among women. According to Poulain, a pattern of female obesity "at the lower end of the social scale" seems to emerge in adolescence, and he highlights a correlative association with higher levels of exercise and associations with slimness in higher socioeconomic adult females and girls over the age of twelve.[55] This emphasis on women, children, and lower socioeconomic status relates directly to the political economy of obesity, in which fatness is conflated with lack of responsibility, and those who are both not fat and in control of the public agenda increase their own status by perpetuating a system of power and rewards that benefit the non-obese over the obese.[56] There is an ongoing process of stigmatization of the obese on multiple levels, arguably aligned to and reinforced by wider neoliberal policies informed by a "methodological individualism" that privileges consumer sovereignty over social pressures when explaining individual food choices. In a typical example, a newspaper report on a research study into the impact of childhood behavior on adult health with the headline "Irresponsible, Careless Kids at Obesity Risk" suggests a direct correlation between "childhood conscientiousness (organized, dependable, self-disciplined) and health status in adulthood."[57]

The dialogue of irresponsibility is particularly interesting when examined in the context of feminist ideas, which are used to both attack and defend female fatness. British psychotherapist and psychoanalyst Susie Orbach's 1978 book *Fat Is a Feminist Issue* points out that "fat is a social disease, and fat is a feminist issue. Fat is *not* about lack of self-control or lack of will power. Fat *is* about protection, sex, nurturance, strength, boundaries, mothering, substance, assertion and rage. It is a response to the inequality of the sexes. Fat expresses experiences of women today in ways that are seldom examined and even more seldom treated."[58] Orbach's research in the 1970s investigated compulsive eating in a white demographic and found that it was a response to powerful social factors that defined particular bodies as embodying particular values. She suggested that imposing these definitions is a common means of oppressing and reducing the status of particular groups—in this case, women in general, not just poorer women. But this is not the general tone of public discourse today. In a typical 2010 press article, Rose Prince suggests that this otherwise powerless group is actually to blame for the crisis, because it associated home cooking with female drudgery and encouraged the development of convenience foods: "It's feminism we have to thank for the spread of fast-food chains and an epidemic of childhood obesity."

Naturally, these arguments reflect their culture. Female body shapes scorned by one culture may be the most desirable for another, although many of these instances are also related to gender-based power structures within those cultures. The Efik and Ibibio "Calabar" peoples of the Nigerian Cross Rivers state practice *Mbodi,* a process by which prospective brides are isolated for weeks or months and fattened to increase their fertility and for their husbands' pleasure (as well as, traditionally, being circumcised). A recent EbonyLife reality television series, *The Fattening Room,* used the idea of a fattening room for a competitive show in which six young women from different African countries were "trained in the arts of womanhood," including "proper" dress, crafts, and cooking.[59] Other societies in which "the capacity to store fat in one's own body is seen as a sign of good health and vitality . . . and individuals displaying a high level of adiposity attain social positions of power and prestige" include Mauritanian and Maori groups.[60] In the late 1990s, Puerto Rican women in Philadelphia were found to aspire to a body that might seem obese to others "for its positive connotations of 'tranquility,' health and lack of problems in life."[61] Likewise, in rural Jamaica, women try to fatten themselves and those they care for "because the plump body is healthily dilated with vital lifeblood and because it suggests to others that one is kind, sociable, and happy to fulfill obligations to kin and community."[62] The *Los Angeles Times* reported on the cultural factors attached to obesity in 2007, suggesting that 78 percent of African American females were overweight or obese, while for all genders, 40 percent of African Americans, 34 percent of Hispanics and 29 percent of Caucasians were obese.[63] This is such a significant proportion of the population that it is hardly surprising that pro-obesity lobby groups such as the National Association to Advance Fat Acceptance (NAAFA), founded in the United States in 1969, or Allegro-Fortissimo in France, founded in the late 1980s, are growing their memberships. Nonetheless, the class-based assumptions remain, and sociological observations find that "lean women are directed to the higher end of society, and heavier ones, to the lower."[64]

PUBLIC HEALTH AND FOOD POLICY

"By training citizens in frugality by means of simple dietary laws . . . how many ills would be prevented," declared François Xavier Lanthenas.[65] Referring to this eighteenth-century doctor and legislator, Michel Foucault suggests that the first task of a doctor is political, and "the struggle against disease must begin with a war against bad government."[66] In line with some of the problematics raised around the notion of biopolitics in chapter 7, the question that seems most difficult to answer is, which government is answerable? Is the health of our own bodies, including their position on an obesity continuum, purely a matter for self-government? Do government legislators have a role and indeed a responsibility in the matter? Or is the answer most likely to be some combination of the two? Historically, diet has always been a core component in individual health and medical treatment, while in the modern period, national governments have consistently applied nutrition-related polices in association with public health campaigns. Given the costs associated with the rise of obesity, these have generally developed into quite detailed guidelines on how much of which foods individuals should consume in order to be "healthy," delivered in the form of admonitions to individuals to exercise personal responsibility. Over the course of the later twentieth century in the United States, these policies moved from the so-called new nutrition of the postwar period, which directed the population to eat sufficient numbers of nutrients by sorting foods into basic food groups and stipulating that people should eat one from each of them each day, to the "negative nutrition" of the 1970s, which focused less on which nutrients to eat and more on which particular foods to avoid.[67] Negative nutrition was endorsed by the U.S. Senate in 1977 and became the core of the United States' nutrition policy. According to Harvey Levenstein, this incentivized food producers to develop and market low-fat, nonfat, cholesterol-free, low-calorie, and other apparently healthier foods. Once advertising standards were

relaxed in the early 1980s (under President Ronald Reagan) and manufacturers were permitted to include health claims such as "heart-wise"
or "health-wise" in their marketing materials and packaging, "products
[not a neutral third party] became the public's prime sources of information about the Negative Nutrition."[68] Although, subsequently, Presi-
dents Bush and Clinton introduced more restrictions, the food industry
exercised its lobbying power on published government dietary advice,
and "the end result reflected many of the continuing paradoxes of
American abundance. While people seem to be adequately informed,
and in receipt of the right messages about dietary intake and exercise,
wider society seems to continue to grow fatter."[69]

Analysis of the global situation is extremely complex, and the pleth-
ora of sources, standards, and reports on dietary advice and its effective-
ness make a detailed overview impossible to accomplish here. However,
it is possible to take a brief overview of a few key countries and their
dietary guidelines and begin to consider what these might mean for the
populations affected by the increasing obesity rates reported by the
WHO. Besides the grounds for skepticism discussed earlier in the chapter, it is far from proven that these kinds of guidelines have any direct
impact on the behavior of populations. Indeed, the U.K. Department of
Environment, Food, and Rural Affairs (DEFRA) reported that while
food spending in lower-income households increased by 1.5 percent
between 2007 and 2011, purchases of fruits and vegetables were 10 percent lower for this group, 22 percent lower for households in income
decile 2, and 15 percent lower for households in income decile 1. This
makes particularly dismal reading in the context of the 2007 Food
Standards Agency report that suggested British people were already eating less than half of the recommended amount of fruits and vegetables,
exceeding the limits for sugar and saturated fats, not reaching the targets
for non-starch polysaccharides, and suffering from a shortage of iron,
folate, and vitamin D.[70] Given that all of these areas are covered by
guidelines, and one of the key messages in government dietary advice
was to eat "five a day" of fruits and vegetables, this would seem to imply
that such advice is, at best, ineffectual.

The comparisons in the following table are drawn from published
guidelines, readily available online to anyone who wishes to consult
them. This small summary sample cannot offer any definitive insights,
but it calls further into question the value and meaning of such guidelines and the difficulties of interpreting their impact and results. On the

face of it, with the exception of Australia, there seems to be a correlation between obesity levels and higher fruit consumption (or a failure to distinguish between fruit and vegetables as distinct categories with their own limits). Furthermore, it might appear that higher calorie limits seem to give permission to overeat or may lead individuals to view the limit as a target to be reached. Brazil jumps out as a particularly interesting example for further study, as it publishes no specific consumption targets but distinguishes foods by their degree of processing and advises accordingly. Brazilians are advised to focus on eating natural foods, limit the intake of processed foods, and avoid ultra-processed foods. The documentation is explicit about which types of foods fall into which category, and it is left to the consumer to decide what exactly to eat, how, and when. It is difficult to imagine some other governments facing up to their food industry in this manner, but it might be that this strategy will gain traction if it proves effective.

As we have seen throughout the chapter, our self-conception is closely tied to our own and others' perceptions of our bodies. This includes, in Warren Belasco's words, several clear definitions of the proper kinds of modern body for different times and places:[71] the "efficient" body (streamlined in the modernist early twentieth century); curvy bodies for baby boomers; skinny bodies for the counterculture; the larger but competitively dieting bodies of the late capitalist era; the "authentic" body sought by consumers of natural, slow, organic, and ethnic foods; the "busy" bodies served by fast foods; and the "responsible" bodies seeking an ethical approach that could "reconcile the material demands of the body with the noble longings of the soul."[72] In the midst of all our daily striving for self-identity, bodily health, long life, and happiness— or even just a decent meal—we are forced to negotiate multiple competing discourses masquerading as choices. We do so in a sociopolitical food environment over which we have less control than the advice we are given might lead us to believe.

Modern social theory, particularly in its combined deployment of social psychology, public policy, and human geography, has been instrumental in analyzing these complex interactions between our embodied selves and the power of social forces exercised by states and markets. Without the insights and vocabulary of modern thinkers— whether it is Brillat-Savarin's invitation to think of gastronomy as a science or more recent feminist and anti-racist critiques of the medicalization of fatness—it is difficult to imagine a discussion of food choices

TABLE 2   DIETARY GUIDELINES: A SAMPLE

Adult targets (over age of 18)	U.S.A.: "My plate My Goals"	Australia: "Australian Guide to Healthy Eating"	U.K.: "Eatwell Guide"	France: "Manger Bouger"	Netherlands: "Schijf van Vijf"	Brazil: "Ten Steps to Healthy Diets" [no specific guide]	China: "Pagoda"	Japan: "Spinning Top"	India: "The Indian Food Pyramid"
% Obesity (M)	34	28	26	22	18	17	7	4	2
% Obesity (F)	35	28	28	22	19	24	8	3	5
Calories (M)	2,000–3,000	Not specified	2,500	Not specified	2790–2200 (decreasing with age)	Not specified	Not specified	2,650	Not specified
Calories (F)	1,600–2,400	Not specified	2,000	Not specified	1750–2020 (decreasing with age)	Not specified	Not specified	1,950	Not specified
% Fruit	5	5	39	40	15		35 (incl. potatoes)	5	
% Vegetables	50	35	(5 a day)	(5 a day)	25	No targets. Advised to focus on "natural" foods, limit "processed" foods, avoid "ultra-processed" foods		35	50–60
% Carbohydrate/Starch/Grains	20 (of which ½ whole grains)	40	37 (incl. potatoes)	30 (1 every meal)	40		40	40	
% Protein	20	15	12	15 (1 or 2 a day)	12		15	17	10–15
% Dairy and alternatives (incl. soy)	10	10	8	20 (3 a day)	8		10	4	
% Oils/unsaturated fats	n/a	Small amounts	1	Limited	Limited	Small amounts	n/a	n/a	20–30
Sugar and saturated fats	0 ("empty calories")	Sometimes	3 (limited)	5 (limited)			Trace	Moderate	0

NOTE: Compiled by Jane Levi

and public health so important to an understanding of modern society. By the same token, as this chapter has demonstrated, modern social theory and its associated conceptions of the self, political economy, public discourse, and anomie are indispensable when explaining the development and impact of public policy in shaping the modern food system.

CHAPTER 12

Consumption

Media, the Domestic Economy, and Celebrity Chefs

In 2016 a new British food critic—the Chicken Connoisseur—became the latest online food sensation.[1] Elijah Quashie, a young man appearing even younger than his modest years, dressed in a shirt and tie, a smart bomber jacket, pristine rare trainers, and, occasionally, gloves, posted *The Pengest Munch,* a series of video reviews of fried chicken shops around London. In his own words, he is "a food critic for mandem who care to know what the finest chicken restaurants in London are and where to find them. And a crep fiend too."[2] Readers unfamiliar with urban slang may like to know that *mandem* is one's group of friends (usually male) and *creps* are cool trainers. Now with eighteen episodes under his belt, including a special involving a visit to New York City (episode 15, "New York Spesh"), plus a new series focused on his beloved footwear, Quashie's YouTube channel is followed by nearly six hundred thousand people and, at time of writing, he has garnered between 1.2 and 4 million views for each one of his short films. This aficionado of cheap chicken, with the help of a friend with video skills and his own engagingly hilarious presentation style, at once claims and subverts mainstream ideas about who and what a food critic is, which food is worthy of critique, and who and where the audience for such opinion might be.

An article in the *Guardian* newspaper offered part of the explanation for where the idea for *The Pengest Munch* emerged from and why. Quashie "attributes his unexpected fame to 'the bald one on MasterChef.' Last year, he was idly watching Gregg Wallace[3] judging a plate of food.

"'I wondered to myself, what makes his opinion more valued than anyone else's? Is it because he's been eating more food, so he has an experienced palate? I'm not sure. I thought, no one is doing this for the type of people who eat at chicken shops.'"[4] By offering a critique of mass-market restaurant food, focusing on accepted and readily recognizable judging criteria of value for money, quantity, and quality of seasoning, sauces, and garnishes, the Chicken Connoisseur is almost a one-man cipher for the themes of this chapter on consumption. Through his persona and those of a few celebrity chefs, television stars, and cookery writers we will question what it is that has made and continues to define the consumer of food in the modern sense, and, to some extent, which foods are judged to fit into the ideal paradigm. For when it comes to food, it is clear that in the twenty-first-century societies of the Global North, we are several types of consumer at the same time: humans that eat to live, individuals that live to eat, and members of various groups that enable us to publicly construct and perform these identities through food. In westernized consumer societies, more privileged socioeconomic groups inhabit a world of gastronomes (self-defined or not) who spend money on particular foods from particular suppliers; develop specific cooking and eating expertise; think, talk, present, and write about their experiences; perform food-specific gender roles; and construct whole identities around particular versions of these ideas. In parallel, the poor are expected to be uncomplaining recipients of whatever their local food bank chooses to supply them with, and at the same time, they are criticized for their unsophisticated tastes, lack of expertise in shopping and cookery, and the bodies that result from the cheap processed foods they are allocated in the gastronomic hierarchy. The Chicken Connoisseur not only challenges ideas about who is qualified to offer their expert opinion on our dinner, but also which meals are worthy recipients of this kind of critique.

To explore some of these questions, this chapter examines food in the context of the contemporary and individualistic notions of personal identity, lifestyle, entertainment, and consumer choice that are largely focused on elite consumers. Taking a step back from the immediate impact of personalities like the Chicken Connoisseur, we discuss Guy Debord's concept of society as representational "spectacle," as a model for considering food media and consumption more broadly. The Debordian notion of the "surrogacy" of experience reveals itself in television food programming and media coverage of celebrity chefs and in the public presentation of fashionable, fetishized dishes, ingredients, and techniques. With reference to Veblen's *Theory of the Leisure Class,* we consider how contemporary societies' ideas about work, leisure, gender, and class are highlighted in the various subtle and not-so-subtle ways that the relative social statuses of cooks, chefs, diners, and critics are represented in media and advertising. A reflection on food-related power relations, as expressed by roles and activity in the domestic sphere, moves the discussion from public to private, revealing the pervasiveness of consumer culture even in the simplest of choices at home and abroad about what we eat, when, and how. It moreover brings to the fore once again the tensions between self and other, change and stratification, and the cultural and the natural that are present throughout our study of food, politics, and society.

Chapter 9 discussed the importance of the "civilizing process"— as laid out by Norbert Elias and developed by Stephen Mennell in his work on manners—to an understanding of the social rules and societal distinctions that the food on our tables, in our larders, and pictured in the cookbooks on our shelves reflects. The "rules" surrounding food have been and continue to be subtle yet powerful indicators of who we are and where we belong. In considering Bourdieu's concept of "cultural capital" with regard to the contemporary food scene, chapter 9 went on to propose that the contemporary foodie or gastronome deploys food as a core element of his or her cultural capital. This notion was developed powerfully by Peter Naccarato and Kathleen LeBesco into a new food-specific model, "culinary capital," notably in their 2012 book of the same name. These ideas of culinary capital manifest in various ways, discussed by numerous other scholars in recent years. Signe Hansen suggested in 2008 that it is the consumer of the output that is the real product in the world of the celebrity chef,[5] while Emma-Jayne Abbott's 2015 study of the impact of celebrity chefs in the home highlights the ways in which the apparent intimacy many of these individuals foster in their books, television performances, and web presence masks their commercial impetus.[6] Naccarato and LeBesco emphasize that as contemporary consumers, we barely even develop our own culinary capital but "bask" in the culinary capital of others, whether by buying pre-prepared (or partially prepared) foods, responding to advertising, or consuming cooking programs, cookbooks, and food memoirs by a select group of celebrity cooks and chefs, as well as, more recently, a certain prescribed degree of diving down-market into the "ordinary" or trashier "dirty" foods that the nineteenth-century gastronome was conditioned to despise.[7] Food remains a crucial element in social distinction, and food choices (which, as we discussed in chapter 11, are constrained by our cultural and economic capital) are used by elite groups to claim significant culinary capital that enables them to "distinguish themselves from the hoi polloi who, with their rushed fast food habits and their woeful ignorance of quality, are unmistakably inferior."[8]

In chapter 11 we touched on gastronomy as an aspect of the construction of the self in relation to food. Here we elaborate on the identity of gastronomes as public, visible, and, above all, consumerist figures whose eating habits are elevated into becoming the very substance of themselves as social beings and who own significant culinary capital. To be a gastronome is to be identified and judged by one's food shopping skills, cookery expertise, restaurant experiences, and gastronomic book collections. The contemporary gastronome, like the nineteenth-century model we met in chapter 11 in the person of Brillat-Savarin, is a figure who wants to be—and, in particular, wants to appear to be—a knowledgeable and respected consumer, a person who understands where and how to acquire the necessary food-related knowledge, and how to exercise it publicly. This close identification between modern gastronomy and individual consumption—in both the narrow and wider senses of the word—has been present from its origins at the end of the eighteenth century. As it developed food into a new arena for self-consciousness, nineteenth-century gastronomy codified, taught, and promoted food knowledge in ways that gave the individual the means to feel confident about and included in these fashionable ideas about eating.

Following the time-honored model of the market economy, having created an uncertainty in its audience, gastronomy set about providing its remediation through public discourse. From the earliest periods, the dissemination of gastronomic knowledge was literary, as books, newspapers, magazines, and periodicals were (and still are) produced in profusion to meet the constructed need for instruction. From Grimod de La Reynière's *Almanach des Gourmands*—an authoritative guide to seasonal produce, fine purveyors, and eating in Paris that was issued in eight revised editions between 1803 and 1812—to the "gastronomical tests" proposed in Brillat-Savarin's *Physiologie du Goût* (1826), one of the key appetites created was for status supported by access to definitive information. Grimod de La Reynière's "juries" tasted and judged everything worth tasting and judging so that readers could shop and give dinners safe in the knowledge that they were buying and serving "the best," while Brillat-Savarin's readers could reassure themselves by making sure they measured up to his "tests." In the twentieth and

FIGURE 10. "Séance d'un Jury de Gourmands Dégustateurs,"
frontispiece to Grimod de la Reynière's *Almanach des Gourmands*,
year 3, 1806. Bibliothèque nationale de France

twenty-first centuries, new media—television, radio, the internet, and
social media—are used in similar ways, with contemporary gastronomic
authorities (writers, critics, chefs, media personalities) exploiting their
wide reach and alternative structures to reach deeper into our everyday
experience. The search for old knowledge and the latest innovations
coexist in this competitive quest for gastronomic ascendancy.

The tests proposed by Brillat-Savarin are typical of the appeal to class identity (or snobbery) that were and remain a key component of this kind of writing.[9] He positions this "meditation" (number 13) as a means of identifying and eliminating potential dinner guests who "do not deserve to have treasures lavished on them, the worth of which they cannot appreciate."[10] The basic premise of the tests, which are given first in Latin, is that the faces of guests should be watched closely while generous quantities of specific dishes of "recognized savor and such indisputable excellence that the mere sight of them must arouse all the gustative powers of a properly constituted man" are served.[11] If the required response is not observed, then the guest is deemed to be an unworthy recipient of these pleasures of the table. The tests are relative to social class, and Brillat-Savarin lays out a series of menus of five dishes for the lower income of "competence," six dishes for the middling income of "affluence," and eight for the high income of "wealth." In the middle and higher income brackets, the menus become wider ranging and more elaborate, and the dishes are made with increasingly rare and expensive ingredients.[12] The expectation is that as one ascends in society, one's tastes and responses should become ever more refined. Equally, one's ability to recognize excellence and show appreciation can mark out those who might deserve to ascend the scale. Brillat-Savarin expects a worthy diner, faced with a truffle-stuffed cockerel and a colossal foie gras as part of their second course, to show "the glow of desire, the ecstasy of enjoyment, and the perfect calm of utter bliss,"[13] while the lower classes are simply expected to appreciate the look of their turkey with chestnuts and sauerkraut and tuck in with a hearty appetite. Taste must be developed, learned, purchased, and displayed within one's proper sphere. On contemporary websites like Chowhound, it is, as LeBesco points out, equally acceptable to show disdain for the lower classes and their tastes. The commentary may be masked by reference to concerns about the health of the large, unattractive bodies on display in chain restaurants and lower-class diners, but the real distinction is socioeconomic.[14]

Thorstein Veblen (1857–1929), an American sociologist and economist of the progressive era, argued that the desire for visible "differentiation in consumption"[15] can be seen in all human societies, including the very earliest, although, and as we discussed in chapter 4, social differentiation in preindustrial periods was based as much on ritual or ceremony as wealth. Focusing on the modern period, Veblen proposed that the current manifestation of this human desire for visible distinction between social groups was based on conspicuous consumption, the acquisition and display of expensive goods and services.[16] This means of making one's superiority visible to others was, for Veblen, a requirement for those wishing to be perceived as operating successfully within modern societies. Just as "conspicuous leisure"[17] protected the status of the nonworking wealthy, the conspicuous consumption of desirable, costly objects that one had not produced oneself would confer significant cultural capital and give the impression of social success, even superiority, to those individuals—mainly men—seen to be consuming them. In industrial societies, class lines are further delineated by ideas about who ought to consume what: "The base, industrious class should consume only what may be necessary to their subsistence. In the nature of things, luxuries and the comforts of life belong to the leisure class," including specific food and drink.[18] Such judgments can clearly be seen to be at play in numerous moralizing commentaries of the nineteenth century suggesting, for example, that the poor should be content with basic food like rough brown bread and not impoverish themselves further by wasting their resources on luxuries like white bread—an attitude prevalent today in reverse, in a world that privileges the chewy, sour, whole-grain artisanal loaf (often made with "heritage" grains) over the packaged loaf of soft white bread from the supermarket, regardless of (or even denying in the face of the scientific evidence) the broadly equivalent nutritional profile.

It is not difficult to see how the gastronome could be reconstructed as an exemplar and conveyor of these ideas. Veblen describes an elite social being whose conspicuous consumption has developed beyond acquiring more than meets basic necessity and has moved into an extreme degree of "specialization as regards the quality of the goods consumed. He consumes freely and of the best, in food, drink, narcotics, shelter, services, ornaments."[19] This activity is as much about public display as it is about the pursuit of personal health and happiness. "Since the consumption of these more excellent goods is an evidence of wealth, it becomes honorific; and conversely, the failure to consume in due quantity and quality becomes a mark of demerit."[20] In order to continue his pursuit of status, the gastronome must develop increasing connoisseurship and good taste and demonstrate it in the public sphere—it is important that his status is seen. The relationship between the host and the guest is proposed as a mutually beneficial arrangement in which the guest is not only a recipient of convivial recreation but also a means to an end for the host. The host's reputation (as a gastronome) is cemented by the guest's performance of the part of vicarious consumer and witness of the host's good taste and access to excess.[21]

While Veblen's work emphasized the habits of the bourgeoisie and upper classes of his own time, contemporary capitalism increasingly depends on mass involvement in the consumerist project. In this context, Alan Warde points out that "symbolic differentiation can also be achieved through the wielding and manipulating of symbols that are not deemed exclusive simply by their cost, but by their association with the good taste of knowledgeable or influential social groups."[22] The increasing value placed on the authentic, the artisanal, and the preservation and elevation of so-called peasant cuisine in the early twenty-first century can be clearly seen to fall into this category. Plotkin and Tilman suggest that in his discussion of the hidden power structures of consumer societies, Veblen demonstrated that the ruling classes had historically attempted to control "the interior social consciousness" of their subordinates, as well as living off the wealth they produced.[23] Now, they argue, "modern businesses [have] turned advertising and mass media to purposes of social control" in just the ways that Veblen identified,[24] and food is one of the means by which this is achieved. To continue our gastronomic example, while only a few can afford actually to travel to and eat at the world's "best" restaurants as defined by the *Guide Michelin* or other prestigious authorities, there are many other levels of participation open to those who wish to associate themselves with this kind of food connoisseurship. Cookbooks, documentary television programs, newspaper articles, chef biographies, Instagram posts, even art and museum exhibitions all serve to make accessible and widen the reach of the symbolic versions of these phenomena, while simultaneously increasing their prestige.

A materialist interpretation suggests that it is the world around us that shapes and limits our possibilities; the place we are born into is largely where we stay, without exercising considerable individual effort to form and manage our own identities.[25] In this context, the digital world is often presented as a democratizing force that has the power to change the automatic and inevitable placement of each of us in our environment; it can apparently supply the means by which anyone can transcend the social relations they were born to. In the case of food (as much as anything else) on the internet, we must question the extent to which this is true. Besides the power relations that govern the degree to which such material is accessible in the first place, Plotkin and Tilman suggest that in the digital realm, as elsewhere, we remain "acutely vulnerable to machinations of fraud and force."[26] These include those exercised upon us by advertising, mass marketing, product placement, and so on, reflecting the Canadian radical thinker C. B. Macpherson's idea that most of the modern individual's engagement with "society" consists "of a series of market relations."[27] In this reading, our engagement with celebrity chefs and gastronomic pundits may feel like membership of a society of like-minded food enthusiasts, but in reality, it is a reflection of the successful manipulation of our tastes and preferences by those with something to sell us. We allow our behavior and habits to be shaped by the pronouncements of selected authoritative voices, while utilizing interactive tools that encourage us to believe that we are exercising our freedom of choice or that our individual input is inherently valuable.[28] We write and check online reviews of restaurants and recipes to validate the tastes and opinions formed by such material in the first place.

Use of image and performance is key to the delivery and promotion of culinary capital to the widest possible audience, and the work of Guy Debord (1931–94) is a helpful structure through which to consider this aspect of consumption. A leading Marxist and founding member of the Situationist International, Debord was thought to be one of the major influences on the May 1968 protests in Paris. It is certainly true that his book, *The Society of the Spectacle* (1967)[29] and the ideas of the wider Situationist International helped to define the social and political movements of the time, as well as offering a framework through which we can interpret contemporary movements. In Debord's analysis of twentieth-century capitalism and consumerism, authentic social relations have been replaced with their representations: the world as we experience it is nothing more than a vast accumulation of images. Providing a bleaker and perhaps radical interpretation of some of the ideas about signs, signifiers, and cultural symbols discussed in chapter 4, Debord presents us with "a society where communication exists in the form of a cascade of hierarchic signals."[30] Managed by the powerful—corporations, politicians, media barons—this image-led spectacle is both the ideology and the experience of late capitalism. In Debord's view, this gives rise to a version of false consciousness in which there is no direct or genuine communication between individuals: everything is experienced through, mediated by, or even manipulated by the spectacle. This kind of critique has been applied to numerous emblematic tropes of contemporary culture, such as Beaudrillard's writing on Biosphere II and theme parks as both expressions of and constructions of contemporary culture.[31] In our current food-obsessed age of television shows, newspaper columns, celebrity chefs, and fashion-led ingredient use, it seems to apply particularly aptly to food and what British novelist Will Self described—in an article paying homage to the currency of Debord's ideas—as "the ludic element of consumer society."[32] Indeed, television cooking programs in particular can be read as perfect examples of the kind of surrogacy typical of the society of the spectacle. Taking a Debordian view, such shows deliver the kind of fabricated, second-hand experiences sold as commodity that are also promoted to consumers in gardening and DIY programs or fly-on-the-wall reality television. In his *Comments on the Society of the Spectacle* (1988), Debord elaborates the discussion using food as a specific example, suggesting that in a world where "a chef will philosophize on cookery techniques as if they were landmarks in universal history,"[33] the spectacle has, in a sense, achieved a revolutionary end to the division of labor.

Veblen emphasized the consistent use of luxuries (including food, drink, and narcotics) as a mode of gender differentiation, declaring that in "the patriarchal regime it has been the office of women to prepare and administer these luxuries, and it has been the perquisite of the men of gentle birth and breeding to consume them."[34] This gendered view of what makes for acceptable consumption extends equally in the modern era to food production. Indeed, it is a truism, bordering on cliché, to say that, when it comes to matters of the kitchen, the division of labor and the valuation placed on that labor has almost entirely favored men. This is apparent in the long-term use of language associated with food-related roles—chefs, until recently, being almost invariably assumed to be male professionals, while cooks were thought of as less professional and much more likely to be women or extremely low-paid, inexpert men. The distinctions made between male and female work are particularly prevalent in the domestic sphere. According to Naccarato and LeBesco, in varied households of similar socioeconomic status, it is "feeding work that creates class identities as well as gender, ethnic, racial and sexual identities."[35] This is not to say that food-related identities are entirely constructed within the home, but the domestic sphere is subject to as many of the external influences as other aspects of our lives are. Even though the mid- to late twentieth-century advertising model of the apron-wearing ur-housewife smilingly serving food to her family has ceased to form part of most people's real-life experience, she still seems reside somewhere in the Western cultural imagination. Today she might be seen returning from work and praised for her success in juggling work and family commitments, even though the construction of such an identity depends to a large extent on the (often invisible, usually low-paid) work of others (often poorer women) outside the home.[36]

Advertisers and food companies who place women in their various food purchasing and serving scenarios often "offer culinary capital as both the reward for utilizing their services and the means of protecting the [American] family."[37] As Naccataro and LeBesco put it, culinary capital delivered in the guise of e-grocers and preprepared foods for home consumption "aids consumer citizens in projecting a better, more discerning and caring self, one who fulfills familial obligations with panache despite the challenges of our fast-paced contemporary lives."[38] Convenience food marketing, delivered to consumers in numerous forms, plays, in particular, on the pursuit of the status conferred by being (and appearing to be) a "good" consumer in whichever context is the most desirable. The woman who can afford the luxuries that allow her to effectively manage work and family regularly appears in food and domestic-appliance related advertising, while her manly partner is more likely to be seen wielding machine tools, driving cars, or drinking neat spirits. When men do appear in this fictional domestic sphere, they often do so in the role of expert chef, who offers to ease the burden of the housewife by providing a technical short cut (such as a Marco-Pierre White Knorr stock cube or a Heston Blumenthal ready-meal for upscale supermarkets). Alternatively, they might perform the role of the summer barbecuing dad (such as in Worcestershire sauce or outdoor grill commercials), or appear as laddish young men presenting quick and easy short cuts to other young men (in the mode of Jamie Oliver).

Many contemporary female television cooks appear to simultaneously adhere to and parody a retrograde version of feminine domesticity, prime examples being the global brands Martha Stewart (from the United States) and Nigella Lawson (of the United Kingdom). Nigella, in particular, is famous for her exploitation of her femininity. The impact on her audience, particularly her male audience, of her sultry style is amusingly portrayed in a Thanksgiving special episode of television comedy *Modern Family*, in which Phil uses the guidance of a "Nigella app" to cook dinner and falls under the spell of her seductive tones. He confesses to camera that he had to pull over to calm down when listening in the car to her instructions on "whipping and beating" the eggs for a meringue, and he guiltily jumps away from the turkey when various members of his family catch him sensuously massaging it as Nigella purrs to "rub the breast and thighs with olive oil."[39] This presentation reflects the carefully constructed persona that Lawson has built over years of writing and television appearances. She implies a carefree carelessness in the kitchen and claims to be no "deranged superwoman" while preparing dinner-party dishes over breakfast wearing figure-hugging clothes.[40] In her *Nigella Express* series, she memorably arrives home after a purported night out drinking with friends, skipping straight from taxi to kitchen to whip herself up a Caramel Croissant Pudding, which bakes as she changes into a black silk dressing gown, before devouring the pudding from the cooking bowl while standing in the kitchen.[41] She is clearly a successful businesswoman, but her persona denies any hint of masculine ambition, and she claims the kitchen deploying all the traditional tropes of domestic femininity. While her knowing self-parody—such as her appearance in *Modern Family* and the name of her baking book, *How to Be a Domestic Goddess,*—affirm Lawson as a clever, compelling personality, her television shows certainly exemplify the glossy spectacle of an impossible yet tightly socially codified life. As Naccataro and LeBesco put it, Lawson, along with other Food Network stars like Paula Deen and Bobby Flay, offers "consumers a means of earning culinary capital through credible performances of a range of gender and class ideologies."[42]

Martha Rosler's artwork *Semiotics of the Kitchen* (1975)[43] predates the Nigella Lawson phenomenon by two decades but continues to offer an alternative critique of the world she inhabits. Rosler presents a direct, first-wave feminist response to traditional portrayals of domestic inequality, representing the home kitchen and its various appliances as definitive symbols of oppression. Standing before the camera at a kitchen table, wearing an apron, the artist moves through the alphabet, naming and wielding an appropriate tool for each letter and, in the process, according to Rosler, replacing "the domesticated 'meaning' of tools with a lexicon of rage and frustration."[44] Becoming more insistent with each letter (we begin to sense trouble at "C—Chopper" and are feeling decidedly unsettled by the time she stiffly thrusts the Pan of *P* in our faces), Rosler throws imaginary ingredients over her shoulder, bashes her table with a meat tenderizer, and conveys with comic deadpan exactly how furious she feels to be forced into this subjective role, constrained within this limited space.[45] As Charlotte Brunsden puts it, "the semiotics of the kitchen has nothing to do with cooking."[46] Another work by Rosler, *The Art of Cooking,* highlights the distinction between male and female perspectives and assumptions when it comes to appreciating and communicating ideas about food.[47] This imaginary conversation between iconic television cook Julia Child and veteran food columnist Craig Claiborne (pieced together from a mixture of excerpts of their actual work and Rosler's own invention) is a discussion, among other things, of taste and whether food is "art." Her imaginary Craig Claiborne chastises Child—and, by extension, all women food writers—for suggesting there is or could be any art to the domestic productions of un-cheffy, nonprofessional cooks. Referring obliquely to Elizabeth David,[48] he patronizingly asks Child, "Forgive me for observing that you ladies are too accepting, not rigorous enough in what you allow as art. If I recall correctly, it was one of your sex who attempted to lionize French provincial cooking over the classic cuisine. It's much like preferring quilts to Picasso."[49] In Claiborne's construction (and despite Child reminding him that it was a man, Austin de Croze, who brought French provincial cooking into the 1923 Salon d'Automne, winning a resolution that "gastronomy, including cooking and cookery, is officially recognized to be an art, the Ninth Art"[50]), it is only by moving into the urbane, professional, masculine sphere that gastronomy comes into being and becomes worthy of the attribution "art."[51]

While the power relations at play in the kitchen, including the professional kitchen, are not only about gender, it has been a crucial element in the transformation of public perceptions of cooking in the late twentieth and early twenty-first centuries. Old-fashioned professional "American" kitchens, those producing ordinary food for ordinary people rather than haute cuisine for the middle and upper classes, were formerly portrayed as "outside of the American success story."[52] These were "places of filthy physical labor where nothing gleamed. They were run not by chefs but by cooks: fat, sweaty chain-smokers in funny hats and undershirts (the exceptions were imagined as motherly 'ethnic' women running low-price-point 'ethnic' restaurants in 'ethnic' neighborhoods)." According to Gwen Hyman's analysis, this meant that the work of reinventing the kitchen as a viable workplace and cooking as a respectable career path consisted, in large part, in "making it manly."[53] Distinguishing themselves from the homely image of motherly cooks and low-grade kitchen skivvies, the new chefs presented themselves as tough rock-n-roll characters, living hard on booze, drugs, and cigarettes, wielding their knives and ingredients like weapons, a stance exemplified by, among others, chefs like Marco-Pierre White. White famously stormed out of the kitchen to berate diners at a top London restaurant who requested condiments, and he was reputed to be the only chef ever to make the hard-as-nails Gordon Ramsay cry when he trained under him. This image was a crucial component in White's fame, and in his youth, he was rarely pictured without cleaver in hand or cigarette in mouth, long hair tousled messily around a sleeplessly gaunt yet handsome face. The writing of Anthony Bourdain lays equal claim to this dark and manly (if seamy) space for a wider legion of line cooks and sous chefs in his biographical writing.[54] This kind of cooking was about a particularly masculine form of expression, designed to exclude women and to be as far from "feminine" notions of food as comfort and nurture as possible.

It is tempting to speculate that this move into the world of cool by young chefs gave rise to its mirror image: the rock musician turned cook or farmer. London's *Financial Times* reported in January 2017 on Groove Armada's Andy Cato and his heritage grain growing in northern France;[55] Alex James, of 1990s Britpop band Blur, has been as well known in recent decades for his cheesemaking and food columns as for his bass playing; and rapper Coolio has restyled himself the Ghetto Gourmet, producing cookbooks and a television show.[56] While some women have made a similar transition, notably the singer Kelis, who appeared at a pop-up restaurant as chef in London in the summer of 2016 and aspires to have her own farm in order to "have control over the whole process,"[57] it is interesting to note how many more men than women appear to be placing themselves in these newly overlapping public worlds. It is also difficult to escape the gendered language used to discuss the phenomenon. The early reports of Kelis's new-found interest seemed unable to take her change of career direction seriously, characterizing her not as a newly formed culinary professional but as a domesticated female "turning into her mum."[58] This is in stark contrast to the rap-appropriate portrayal of Coolio as a chef, surrounded by glamorous female kitchen assistants, or the self-styled "Vegan Black Metal Chef" Brian Manowitz of the band Forever Dawn growling vegan curry recipes over an industrial metal soundtrack.[59]

Another way to make cooking appear more manly is to make it more scientific, more analytical, and the rise of Molecular Gastronomy elevated the embrace of futuristic techniques, equipment, and ingredients to high fashion. For many chefs involved with this movement, at least in the public imagination, the association with science went a little too far. Its 2006 manifesto, written and signed by international luminaries of the style, including the chefs Ferran Adrià, Heston Blumenthal, and Thomas Keller, as well as renowned food history and science writer Harold McGee, was designed to clear up apparent misunderstandings about their relationship with science, recasting the group as more of a creative movement rooted in tradition. Their declared commitment to excellence with integrity; tradition and craft; innovation, especially in technique; and collaboration, in every sense, to realize the expressive potential of food was a plea for the creative respect of other chefs and diners who might have misunderstood the gimmickry of foams and emulsions when in the hands of an inadequate chef.[60] Nonetheless, Heston Blumenthal, recast as a "multi-sensory" chef, usually appears in pristine chef's whites and large glasses, almost chemistry-lab protective goggles, taking on the air of a boffin able to deconstruct a dish into artful components while at the same time explaining the scientific processes lying beneath the transformation of the ingredients. As he says in his series for the Discovery Channel, *Kitchen Chemistry,* "a kitchen is a bit like a science lab."[61] The approach has been rewarded not only with Michelin star ratings for his restaurants but also by being named as one of the Royal Society of Chemistry's "175 Faces of Chemistry" in 2016.[62] In a similar vein, the eye-wateringly expensive and lavishly produced six-volume *Modernist Cuisine,* a cookbook project funded by Microsoft millionaire Nathan Myhrvold, also adopts a deconstructivist, scientific approach. Illustrated with cutaway photographs revealing what is actually happening inside the oven or pot, *Modernist Cuisine* discusses in detail the scientific processes involved in cooking and includes recipes from important "molecular" chefs around the world.[63] It was inducted in the Gourmand Cookbook Awards' Hall of Fame in 2011 as the most important book—in any genre—of the first decade of the century.[64] The online promotional video for it emphasizes its masculine blend of incredible science and lavish creativity.[65] These examples could not be more different in style from the deliberately domestic and often artfully flawed offerings of their female counterparts, such as Nigella Lawson or Ina Garten (the "Barefoot Contessa").

What all of these public figures have in common is the importance, almost above their craft and style of cuisine, of their person and images of it. Like most of the book covers by the personalities discussed here (few of which do not involve a photograph of the author), the cover of the 2012 memoir of nouvelle cuisine, *Mémoires de Chefs,* features a photograph of three of the men who gave rise to that movement. This cover highlights the increasing focus on personality and personal image that developed in the late twentieth century. In their own era, while they were themselves well known, these chefs' food took the visual as well as the sensory center stage, and it was still possible to have heard of and be influenced by them without ever having seen their picture.[66] In the late twentieth and early twenty-first century, though, "chefs began to be people to whom attention was paid. . . . The idea that kitchen work was only for grubby workers, histrionic Europeans, and accent-ridden earth-mother types began to disappear"⁶⁷ and with it, the status of the role rose. A chef became a person who could, if successful, develop valuable cultural capital along with financial capital or who could embellish their existing cultural capital in one field with achievement in another (along the lines of Bill Buford, the *New Yorker* staff writer who relaunched the literary magazine *Granta* in 1979 and who, in the early 2000s, decided to work as Chef Mario Batali's "kitchen bitch" in the upscale Babbo restaurant in New York and then went on to produce a memoir and, later, specialist food writing).⁶⁸

In today's model of connoisseurship, it is the contribution to the spectacle that is most visibly valued. According to Parkhurst Ferguson and Zukin, even gastronomic Parisian restaurants now provide "a star system of new French chefs who are no longer crafts 'workers,' or even 'artists,' in a time-honored production system, but media stars."⁶⁹ This is true well beyond France and applies across the developed world, in which "lifestyle" media has become the primary means by which many people regularly experience cooking and restaurant dining. Public consumption, the act of dining out—and posting on Instagram the photographs of the dishes to prove it—has become one of the most fashionable pursuits of our time, even for those fundamentally uninterested in the art of cooking or even eating. As Veblen would have it, "We readily, and for the most part with utter sincerity, find those things pleasing that are in vogue,"⁷⁰ and it cannot be denied that public consumption of the right food in the right places has become a key marker of our place in society. As much as the houses we live in, the clothes we wear, or the leisure interests we pursue, our food choices betray the kind of consumers we are and aspire to be, as well as the fact that we are, above all, consumers in the modern sense: our cultural capital becomes our social capital, and in the development of a specifically culinary capital, our individual taste has a great deal less to do with it than we would like to think.

Notes

CHAPTER I INTRODUCTION

1. Georg Simmel, "Sociology of the Meal" in *Simmel on Culture: Selected Writings,* ed. D. Frisby and M. Featherstone (London: Sage, 1997), 130.

2. Ibid., 130.

3. Here we are echoing Jack Goody's early identification of the "process of providing and transforming food" as covering "phases of production, distribution, preparation and consumption [and] disposal" in Jack Goody, *Cooking, Cuisine and Class: A Study in Comparative Sociology* (Cambridge: Cambridge University Press, 1982), 37.

4. For reasons of both space and expertise, we deal in this book mainly with western, and indeed Anglophone literatures on food, politics, and society. An account with a more Francophone flavor can be found in the recent English translation of Jean-Pierre Poulain's *The Sociology of Food: Eating and the Place of Food in Society,* trans. Augusta Dörr (London: Bloomsbury, 2017).

5. Good recent overviews can be found in Ken Albala, ed., *Routledge International Handbook of Food Studies* (London: Routledge, 2014); Carol Counihan and Patricia van Esterik, eds., *Food and Culture: A Reader,* 2nd ed. (London: Routledge, 2008); and R.J. Herring, *The Oxford Handbook of Food, Politics, and Society* (Oxford: Oxford University Press, 2015).

6. With apologies to Charles Tilly, *Big Structures, Large Processes, Huge Comparisons* (New York: Russell Sage Foundation, 1989).

7. Stephen Mennell, Anne Murcott, and Anneke H. van Otterloo, *The Sociology of Food: Eating, Diet and Culture* (London: Sage, 1992), 14.

8. Bruce Mazlish, *A New Science: The Breakdown of Connections and the Birth of Sociology* (Oxford: Oxford University Press, 1989).

9. Ibid., 12.

10. Otto Brunner, Werner Conze, and Reinhardt Koselleck, eds., *Geschichtliche Grundbegriffe: Historisches Lexikon zur politisch-sozialen Sprache in Deutschland* (Stuttgart: Klett-Cotta, 1972).

11. J. G. A. Pocock, *The Machiavellian Moment: Florentine Political Thought and the Atlantic Republican Tradition* (Princeton, NJ: Princeton University Press, 1979); Quentin Skinner, *The Foundations of Modern Political Thought,* 2 vols. (Cambridge: Cambridge University Press, 1978).

12. Karl Marx and Frederick Engels, *The Communist Manifesto* (London: Verso, 1998), 38.

13. Émile Durkheim, *The Rules of Sociological Method* (1895).

14. Nicos Mouzelis, *Sociological Theory: What Went Wrong? Diagnosis and Remedies* (London; Routledge, 1995), 2–3.

15. Norbert Elias, *The Civilizing Process,* 2nd ed. (Oxford: Wiley-Blackwell, 2000).

16. Alfred W. Crosby Jr., *The Columbian Exchange: Biological and Cultural Consequences of 1492* (Westport, CT: Greenwood, 1972).

17. As chronicled in M. J. O'Brien's evocative *We Shall Not Be Moved: The Jackson Woolworth's Sit-In and the Movement It Inspired* (Jackson: University of Mississippi Press, 2013).

18. Craig Calhoun, *Habermas and the Public Sphere* (Cambridge, MA: MIT Press, 1992).

19. As recounted by Sami Zubaida in "Our Kitchen in 1940s Baghdad," Oxford Food Symposium, 2012, http://www.oxfordsymposium.org.uk/wp-content/uploads/2013/06/Zubaida.pdf.

20. Karl Marx, *Capital,* vol. 1 (London: Penguin, 1993 [1867]) 637–38.

21. Ibid., 638.

22. Amartya Sen, *Poverty and Famines: An Essay on Entitlement and Deprivation* (Oxford: Oxford University Press, 1990), 1.

23. Ulrich Beck, *Risk Society: Towards a New Modernity* (London: Sage, 1992).

24. Roland Barthes, "A Psychosociology of Contemporary Food Consumption," in *Food and Drink in History: Selections from the Annales Economies, Sociétés, Civilizations,* ed. Elborg Forster and Orest Raum (Baltimore, MD: Johns Hopkins University Press, 1979), 168.

CHAPTER 2 THE NATURAL AND THE SOCIAL

1. Graeme Barker, *The Agricultural Revolution in Prehistory* (Oxford: Oxford University Press, 2006), 1–9.

2. Jared Diamond, *The Rise and Fall of the Third Chimpanzee* (London: Radius, 1991), 9–20, 27–55.

3. Ibid., 121–49.

4. Richard Wrangham, *Catching Fire: How Cooking Made Us Human* (London: Profile, 2009).

5. Marshall Sahlins, "The Original Affluent Society," in *Stone Age Economics* (London: Tavistock, 1974), 1–40. James C. Scott has recently highlighted the significance of these sedentary wetland populations in a variety of settings,

including southern Mesopotamia, which seem to have come into being several thousand years before the emergence of crop-field agriculture. At the same time, there is considerable evidence of crops being planted and harvested without sedentism: in this scenario, hunter gatherers lived in or near the crop fields only for a small part of the year, retaining their nomadic lifestyle for the remainder. James C. Scott, *Against the Grain: A Deep History of the Earliest States* (New Haven, CT: Yale University Press, 2017), 10–11.

6. Diamond, *Rise and Fall,* 111; Barker, *Agricultural Revolution,* 9–33.

7. Jared Diamond. *Guns, Germs, and Steel: A Short History of Everybody for the Last 13,000 Years* (London: Jonathan Cape, 1997), 119.

8. Ibid., 157–75.

9. Ibid., 167.

10. Ibid., 120.

11. Ibid., 124.

12. Peter Bellwood, *First Farmers: The Origins of Agricultural Societies* (Oxford: Basil Blackwell, 2005), 117.

13. Ibid., 114.

14. Ibid., 114.

15. Ibid., 42–43.

16. Scott, *Against the Grain,* 116–49.

17. Diamond, *Guns, Germs, and Steel,* 136.

18. Ibid, 140–41.

19. Ibid., 181.

20. Ibid., 190.

21. Sahlins, "Original Affluent Society," 15–17.

22. Thomas Hobbes, *Leviathan* (Cambridge: Cambridge University Press 1996), 89.

23. Quoted in Sahlins, "Original Affluent Society," 25.

24. Barker, *Agricultural Revolution,* 396–97; Diamond, *Guns, Germs, and Steel,* 146–56.

25. Diamond, *Guns, Germs, and Steel,* 106–7.

26. Barker, *Agricultural Revolution,* 406–7.

27. Diamond, *Guns, Germs, and Steel,* 106.

28. Ibid., 89.

29. Max Weber, *The Agrarian Sociology of Ancient Civilizations,* trans. R. I. Frank (London: New Left Books, 1976), 70.

30. Scott, *Against the Grain,* 7

31. Diamond, *Guns, Germs, and Steel,* 193–214.

32. Examples can be found in the anthropological classic by Meyer Fortes and Edward Evans-Pritchard, eds., *African Political Systems* (London: International African Institute, 1987). See also Diamond, *Guns, Germs, and Steel,* 274–75.

33. V. Gordon Childe, *Man Makes Himself* (London: Watts, 1936), and *What Happened in History?* (London: Penguin, 1942). These two seminal works were widely influential and, though challenged and amended by later research, remain an important starting point.

34. Karl Wittfogel, *Oriental Despotism: A Comparative Study of Total Power* (New Haven, CT: Yale University Press, 1957).

35. Michael Mann, *The Sources of Social Power* (Cambridge University Press, 1986), 4 vols.
36. Ibid, vol. 1, 75 (italics in original).
37. Jean Bottero, *The Oldest Cuisine in the World: Cooking in Mesopotamia* (Chicago: Chicago University Press, 2004), 11–12.
38. Diamond, *Guns, Germs, and Steel*, 218.
39. Bottero, *Oldest Cuisine*, 16–17.
40. Ibid., 20, 22.
41. Ibid., 36–51.
42. Ibid., 26.
43. Ibid., 89–96.
44. Ibid., 90.
45. Ibid., 92–93.
46. Ibid., 89.
47. Ibid., 94–96.

CHAPTER 3 EXCHANGE

1. Felipe Fernández-Armesto also recounts this story in his *Near a Thousand Tables: A History of Food* (New York: Free Press, 2002).
2. See Sami Zubaida, "A Global Palate," *The Middle East in London* 12 (2016): 11–12; and John Aytom, *The Diner's Dictionary: Word Origins of Food & Drink* (Oxford: Oxford University Press, 2012), 69.
3. Ayto, *The Diner's Dictionary*, 388.
4. Crosby, *The Columbian Exchange*.
5. Gary Paul Nabhan, *Cumin, Camels and Caravans: A Spice Odyssey* (Berkeley: University of California Press, 2014).
6. Here, Eric Wolf's magisterial overview of modern world history is useful in addressing Nabhan's confusion between globalization as any instance of long-distance commerce, and the historically specific mode of production, capitalism, that has encompassed the whole planet: "What we must be clear about" Eric Wolf suggests, "is the analytical distinction between the employment of wealth in the pursuit of further wealth, and capitalism as a qualitatively different mode of committing social labour to the transformation of nature." Eric R. Wolf, *Europe and the People without History* (Berkeley: University of California Press, 1997), 298.
7. For a more detailed definition and discussion of colonialism and empires, see Alejandro Colás, *Empire* (Cambridge: Polity Press, 2007).
8. In chapter 26 of *Capital*, vol. I (London: Penguin, 1972), as discussed in chapter 5.
9. Joseph Schumpeter, *Capitalism, Socialism and Democracy* (London: Routledge, 2010 [1942]).
10. A comprehensive list of food plants of American origin is provided in Nelson Foster and Linda S. Cordell, eds., *Chiles to Chocolate: The Foods the Americas Gave the World* (Tucson: University of Arizona Press, 1992).
11. Rachel Laudan, *Cuisine and Empire: Cooking in World History* (Berkeley: University of California Press, 2013). As Laudan suggests, nixtamalization

(alkali treatment) of maize, making it far more nutritional, was not adopted in the Old World. Nor was the Andean technique of soaking and then freeze-drying potatoes, which yields a long-life product known as *chuño*.

12. See Redcliffe N. Salaman, *The History and Social Influence of the Potato,* 2nd ed. (Cambridge: Cambridge University Press, 2010); and for a good overview, see John Reader, *The Untold History of the Potato* (London: Vintage Books, 2008); and Ellen Messer, "Potatoes (White)," in *Cambridge World History of Food,* ed. Kenneth F. Kiple and Kriemhild Coneè Ornelas, 187–200 (Cambridge: Cambridge University Press, 2000).

13. Fernand Braudel, *The Structures of Everyday Life: Civilization & Capitalism,* vol. 1 (London: Phoenix Press, 1981).

14. See Rebecca Earle, "The Columbian Exchange," in *The Oxford Handbook of Food History,* ed. Jeffrey M. Pilcher, 341–59 (Oxford: Oxford University Press, 2012); James Walvin, *Fruits of Empire: Exotic Produce and British Taste, 1660–1800* (Houndsmills and London: Macmillan Press, 1997); William H. McNeill, "How the Potato Changed the World's History" *Social Research* 66 (1999): 67–83.

15. As reported in Fernández-Armesto, *Near a Thousand Tables,* 179. See also Sucheta Mazumdar, "The Impact of New World Crops on the Diet and Economy of China and India, 1600–1900," in *Food in Global History,* ed. Raymond Grew, 58–78 (Boulder, CO: Westview, 1999).

16. Diamond, *Guns, Germs and Steel,* 183–84; and Crosby, *The Columbian Exchange.*

17. Crosby, *The Columbian Exchange,* 164.

18. Marcy Norton, "Tasting Empire: Chocolate and the European Internalization of Mesoamerican Aesthetics," *American Historical Review* 111 (2006): 661.

19. See Rebecca Earle, *The Body of the Conquistador: Food, Race and the Colonial Experience of Spanish America, 1492–1700* (Cambridge: Cambridge University Press, 2014), 149–55. William Rubel also makes the point in his book *Bread* (London: Reaktion, 2012) that to this day in Mexico, wheat-based bread loaves are sold in *panaderías* (bakeries) while corn tortillas are made and purchased in *tortillerías,* with all the sociological and historical connotations implied in the distinction.

20. McNeill, "How the Potato."

21. John H. Elliott, *Empires of the Atlantic World: Britain and Spain in America, 1492–1830* (New Haven, CT: Yale University Press, 2006), 89.

22. Alfred Crosby Jr., *Ecological Imperialism: The Biological Expansion of Europe, 900–1900,* 2nd ed. (Cambridge: Cambridge University Press, 2004), 3.

23. John C. Super and Luis Alberto Vargas, "Mexico and Highland Central America," in *Cambridge World History of Food,* ed. Kenneth F. Kiple and Kriemhild Coneè Ornelas (Cambridge: Cambridge University Press, 2000).

24. John F. Richards, *The Unending Frontier: An Environmental History of the Early Modern World* (Berkeley: University of California Press, 1997), 361 ff.

25. "Pastoralism," Melville has written, "had no recognizable social counterparts in the preconquest world outside of the Andes, and its introduction

involved not only the addition of exotic species but also a completely alien perception of the natural resources and their use; indeed, it involved the formation of completely new systems of production." Elinor G.K. Melville, *A Plague of Sheep: The Environmental Consequences of the Conquest of Mexico* (Berkeley: University of California Press, 1997), 8.

26. Super and Vargas, "Mexico and Highland Central America."

27. Jeffrey M. Pilcher and Rachel Laudan, "Chiles, Chocolate, and Race in New Spain: Glancing Backward to Spain or Looking Forward to Mexico?" *Eighteenth-Century Life.* 23 (1999): 59–70.

28. Judith A. Carney and Richard Nicholas Rosomoff, *In the Shadow of Slavery: Africa's Botanical Legacy in the Atlantic World* (Berkeley: University of California Press, 2009), 179.

29. Ibid., 180

30. Jayeeta Sharma, "Food and Empire" in *The Oxford Handbook of Food History,* ed. Jeffrey M. Pilcher, 241–57 (Oxford: Oxford University Press, 2012).

31. Donna R. Gabaccia, "Colonial Creoles: The Formation of Tastes in Early America, in *Taste Cultures: Experiencing Food & Drink,* ed. Carolyn Korsmeyer, 79–85 (London: Bloomsbury, 2005), 84.

32. Fernando Ortíz, *Cuban Counterpoint: Tobacco and Sugar* (Durham, NC: Duke University Press, 1995), 102–3.

33. Ibid., 6.

34. Massimo Livi Bacci, *Conquest: The Destruction of the American Indios* (Cambridge: Polity Press, 2008).

35. Crosby, *The Columbian Exchange,* 166.

36. Regional variations and mixed access to records have made it very difficult, even for specialists, to arrive at anything more than estimated figures. See Richards, *Unending Frontier,* 337–40; and Nicolás Sánchez-Albornoz, "The Population of Colonial Spanish America," in *Colonial Spanish America,* ed. Leslie Bethell, 3–36 (Cambridge: Cambridge University Press, 1987).

37. Sidney W. Mintz's classic *Sweetness and Power: The Place of Sugar in Modern History* (London: Penguin, 1986) remains unsurpassed as a comprehensive overview. See also the magisterial synthesis of more recent scholarship on the subject in Robin Blackburn, *The American Crucible: Slavery, Emancipation and Human Rights* (London and New York: Verso, 2011).

38. As Mintz indicates, the mechanized, time-disciplined, rationalized, and alienated division of labor of the sugar plantation meant it should be considered "very early in its career as a form of productive organization . . . an industrial enterprise," *Sweetness and Power,* 50.

39. For an excellent account of this varied literature, including works by the Brazilians Furtado, Dos Santos, and Cardoso, the German-American Frank, or the Guayanese Rodney, see Cristóbal Kay, *Latin American Theories of Development and Underdevelopment* (London: Routledge, 1989).

40. Andre Gunder Frank, "The Development of Underdevelopment," *Monthly Review* 18 (1966): 17–31.

41. Kenneth Pomeranz, *The Great Divergence: China, Europe, and the Making of the Modern World Economy* (Princeton, NJ: Princeton University Press, 2000).

42. A distinction made in different ways by Karl Marx and Karl Polanyi, but elaborated upon most recently by Ellen Meiksins Wood in her *The Origin of Capitalism* (London: Verso, 2002).

43. Jeffrey M. Pilcher, "The Caribbean from 1492 to the Present," in *Cambridge World History of Food*, ed. Kenneth F. Kiple and Kriemhild Conee Ornelas, 1278–87 (Cambridge: Cambridge University Press, 2000). See also Richard Wilk, *Home Cooking in the Global Village: Caribbean Food from the Buccaneers to Ecotourists* (Oxford: Berg, 2006).

44. Sidney W. Mintz, "Caribbean Marketplaces and Caribbean History," *Radical History Review* 27 (1983): 116.

45. Walvin, *Fruits of Empire*, 121.

46. Marcy Norton, *Sacred Gifts, Profane Pleasures: A History of Tobacco and Chocolate in the Atlantic World* (Ithaca, NY: Cornell University Press, 2008), 165.

47. C. A. Bayly, *The Birth of the Modern World, 1780–1914* (Oxford; Blackwell, 2004), 51. See also Jan de Vries, *The Industrious Revolution: Consumer Behaviour and the Household Economy, 1650 to the Present* (Cambridge: Cambridge University Press, 2008); and Troy Bickham, "Eating the Empire: Intersections of Food, Cookery and Imperialism in Eighteenth-Century Britain," Past and Present 198 (2008): 71–109.

48. Harriet Friedmann, "Feeding the Empire: The Pathologies of Globalized Agriculture," in *Socialist Register: Empire Reloaded,* ed. Colin Leys and Leo Panitch (London: Merlin Press, 2005), 125.

49. Though some insist this is just another instance of cultural appropriation by imperialist or neo-colonial societies. See for instance, Lisa Heldke, "Let's Cook Thai: Recipes for Colonialism," in *Food and Culture: A Reader,* ed. Carole Counihan and Penny van Esterik, 2nd ed., 327–41 (New York and London: Routledge, 2008).

50. Jennifer Parker Talwar offers a fascinating applied sociology of New York City's fast food workers, illustrating how social mobility, citizenship, ethnic identity, market segmentation, racism, and gender inequalities all play a critical part in the reproduction of that sector, in her *Fast Food, Fast Track: Immigrants, Big Business, and the American Dream* (Boulder, CO: Westview Press, 2004).

51. http://www.cateringcircle.co.uk/index.html.

52. Igor Cusack, "African Cuisines: Recipes for Nation-Building?" *Journal of African Cultural Studies* 13 (2000): 207–25.

53. See for instance, Nick Cullather, *The Hungry World: America's Cold War Battle Against Poverty in Asia* (Cambridge, MA: Harvard University Press, 2010).

54. See the suggestive essay by Sydney Watts, "Food and the Annales School," in *The Oxford Handbook of Food History,* ed. Jeffrey M. Pilcher, 3–22 (Oxford: Oxford University Press, 2012).

CHAPTER 4 CULTURE

1. Helen Sharman, *Seize the Moment* (London: Gollancz, 1993), 99.

2. See Smith and Christian, *Bread and Salt* (Cambridge: Cambridge University Press, 1984).

3. Anne Murcott demonstrated there is an immense complexity of social, class- and gender-related issues at stake when discussing what might seem to be the simplest food-related questions. See "Cooking and the Cooked: A Note on the Domestic Preparation of Meals," in *The Sociology of Food and Eating,* ed. Anne Murcott (Aldershot: Gower, 1983).

4. See, for example, Paul Atkinson, "Eating Virtue," in *The Sociology of Food and Eating,* ed. Anne Murcott (Aldershot: Gower, 1983).

5. Frederick John Simoons, *Eat Not This Flesh* (Madison: University of Wisconsin Press, 1961), 41. *Mouth* is used here in the sense of eating habits rather than physiognomy.

6. J. G. Frazer, *The Golden Bough* (Oxford: Oxford University Press, 2009).

7. W. R. Smith, *Lectures on the Religion of the Semites,* ed. B. S. Turner (London: Routledge, 1997). Smith was charged with heresy in 1878 as a result of an *Encyclopedia Britannica* article in which he questioned the dating of the laws in Deuteronomy and suggested the Bible was not literally true.

8. Frazer's work on totems was cited by Sigmund Freud, among others.

9. Simoons, *Eat Not This Flesh,* 118.

10. Émile Durkheim, *The Elementary Forms of Religious Life* (New York: Free Press, 1995), 127.

11. Émile Durkheim and Marcel Mauss, *Primitive Classification* (London: Cohen & West, 1963).

12. Bronislaw Malinowski, *Argonauts of the Western Pacific* (London: Routledge, 2002).

13. Claude Lévi-Strauss, *The Raw and the Cooked* (Harmondsworth, England: Penguin Books, 1986), e.g., 334–36; *The Origin of Table Manners,* (New York: Harper Colophon, 1979), 471–508. Also reproduced in "The Culinary Triangle," in *Food and Culture: A Reader,* ed. C. Counihan and P. van Esterik (New York: Routledge, 2008).

14. See the detailed exposition of analogous units which demonstrate that "eating, like talking, is a patterned activity, and the daily menu may be made to yield an analogy with linguistic form," in Mary Douglas, "Deciphering a Meal," *Daedalus* 101, (1972).

15. Lévi-Strauss, *Origin of Table Manners,* 3 and 477–95.

16. Rachel Laudan, afterword in *Food & Foodways* 19, (2011) 161 (our emphasis).

17. "Slow Food is a global, grassroots organization, founded in 1989 to prevent the disappearance of local food cultures and traditions, counteract the rise of fast life and combat people's dwindling interest in the food they eat, where it comes from and how our food choices affect the world around us." From www.slowfood.com.

18. "Starting with a basic definition of Real Bread simply as being made without the use of any so-called 'processing aids' or other artificial additives, the Campaign seeks, finds and shares ways to make bread better for us, better for our communities and better for the planet." From www.sustainweb.org /realbread.

19. This accelerated process for making bread (from mixing to sliced packaged loaf in about three-and-a-half hours) was developed in Chorleywood,

Hertfordshire, England in 1961. Vitamin C (ascorbic acid), fat, and yeast are added to the flour and water, and the dough is worked intensively with high-speed mixers.

20. The term *artisanal* has a specific meaning in contemporary baking that does not prescribe the banishment of machinery or the degree of hand-crafting one might assume. See William Rubel, "Artisan Bread," in *Food and Architecture: At the Table*, ed. Samantha L. Martin-McAuliffe, 151–66 (London: Bloomsbury, 2016).

21. Roland Barthes, *Elements of Semiology* (New York: Hill and Wang, 1977), 27–28.

22. *Empire of Signs* (New York: Hill and Wang, 1983), 11–26.

23. *Mythologies* (London: Vintage, 1993), 58.

24. Mary Douglas, *Implicit Meanings* (London: Routledge, 1999), 232 and 194.

25. Mary Douglas, *Purity and Danger* (London: Routledge, 2002), 51–71.

26. Douglas, *Implicit Meanings*, 311.

27. Durkheim, *Elementary Forms*, 46.

28. Martin Luther, "The Babylonian Captivity of the Church," www.lutherdansk.dk/Web-babylonian%20Captivitate/Martin%20Luther.htm, Project Wittenberg, ed. Robert E. Smith (1522), 2:27.

29. Archbishops' Council, "The Book of Common Prayer," last modified 2014, www.churchofengland.org/prayer-worship/worship/book-of-common-prayer/articles-of-religion.aspx#XXXVIII.

30. The "oral Torah," or Talmud, is the explanation of what these laws mean, comprised of the Mishnah, in which the oral laws were first written down in the second century B.C.E., and the Gemara, completed in the fifth century B.C.E. These are further elaborated on in subsequent commentaries.

31. Leviticus 11:1–47, 17:10–15; Deuteronomy 14:2–21; Exodus 23:19, 34:26.

32. A helpfully simple set of summaries on this and other aspects of Judaism can be found at the Jewish FAQ website. See Tracey R. Rich, "Judaism 101," www.jewfaq.org/index.shtml.

33. Lev. 11:8.

34. Lev. 11:3; Deut. 14:4–8. Camel meat is also rejected by Ethiopian Christians, who see its consumption as a violation of custom (as well as a Muslim trait) that could be punished by excommunication. See Simoons, *Eat Not This Flesh*, vii, 42.

35. Lev. 11:13.

36. Lev. 11:9–12; Deut. 14:9–10.

37. Deut. 14:19; Lev. 11:41–43.

38. Lev. 17:10. Early Christians also observed this prohibition on blood and animals killed by strangling, a tradition continued by the Orthodox Church after the Western Church abandoned it sometime around the fourteenth century. See Smith and Christian, *Bread and Salt*, 12–13.

39. "Thou shalt not seethe a kid in its mother's milk." Ex. 23:19, 34:26; Deut. 14:21. Soler associates this prohibition with the rules against human incest, suggesting that to mix milk and meat would be a kind of "culinary incest."

40. See, for example, Rabbi Hayim Halevy Donin, *To Be a Jew* (New York: Basic Books, 1972), chap. 6, "The Dietary Laws," 97–120.

41. Douglas elaborates on some aspects of this work criticized by other scholars in her essay, "Deciphering a Meal."

42. Douglas, *Purity and Danger*, 63.

43. Ibid., 67. For example, the prohibition against combining meat and dairy products, which can be read as a prohibition against incest.

44. Jean Soler, "Biblical Reasons: The Dietary Rules of the Ancient Hebrews," in *Food: A Culinary History*, ed. J.-L. Flandrin and M. Montanari, trans. A. Sonnenfeld (New York: Columbia University Press, 1999), 50–51.

45. Ibid., 48.

46. "And every creeping thing that creepeth upon the earth shall be an abomination; it shall not be eaten." Lev. 11:41.

47. *The Holy Qur'ān* (Leicester: The Islamic Foundation, 1975).

48. This practice, known as *dhabh*, requires that the animal should be killed with an incision across the throat and the blood drained as fully as possible. There are some that interpret this to mean that the animal may not be stunned before slaughter, though others do allow this as part of ethical modern food production systems. See, for example, Johan Fischer, *The Halal Frontier: Muslim Consumers in a Globalized Market*, ed. Laurel Kendall (New York: Palgrave Macmillan, 2011), 6.

49. "But if any is forced / To hunger, with no inclination / To transgression, God is / Indeed Oft-forgiving, / Most Merciful." Qur'an, 5:3.

50. This gave rise to a fascinating conundrum in the Islamic Republic of Iran after the 1979 revolution. The scaleless sturgeon was forbidden to Shi'a Muslims, threatening Iran's valuable caviar trade; in 1983 evidence of scales at the tail end of the fish was presented to Khomeini and he issued a ruling *(fatwa)* permitting consumption of the fish and its eggs. See Houchang Chehabi, "How Caviar Turned Out to Be Halal," *Gastronomica* 7, (2007): 17–23.

51. The verses citing "wine" (alcohol) as the work of Satan also define gambling as an "abhorrence" (5:90). Tarif Khalidi, *The Qur'an: A New Translation* (London: Viking Penguin, 2008).

52. Halil İnalcık, *The Ottoman Empire* (London: Weidenfeld & Nicolson, 1973).

53. With thanks to Sami Zubaida for this—and many other—apposite jokes.

54. Fischer, *The Halal Frontier*, 9.

55. Douglas, *Purity and Danger*, 51.

56. Core disgust requires that there is a possibility that the object of disgust might be eaten, that it is in some way offensive, and that it either has or we believe it has the power to contaminate us. For a brief summary, see Jane Levi, "Disgust," in *The Sage Encyclopedia of Food Issues*, ed. Ken Albala (Thousand Oaks, CA: Sage, 2015).

57. There is an expansive and fascinating literature on disgust and the emotions. See in particular the work of Paul Rozin, e.g., P. Rozin, J. Haidt, C. McCauley, and S. Imada, "Disgust: Preadaptation and the Cultural Evolution

of a Food-Based Emotion," in *Food Preferences and Taste,* ed. H. Macbeth, 65–82 (Oxford: Berghahn Books, 1997).

58. Veronika E. Grimm, *From Feasting to Fasting, the Evolution of a Sin* (London: Routledge, 1996), 68.

59. Ibid., 160.

60. St. Thomas Aquinas, *Summa Theologica,* vol. 3 (Allen, TX: Christian Classics, 1981), question 148.

61. Marjo Buitelaar, "Living Ramadan: Fasting and Feasting in Morocco," in *The Taste Culture Reader: Experiencing Food and Drink,* ed. Carolyn Korsmeyer (Oxford: Berg, 2005).

62. Carolyn Rouse and Janet Hoskins, "Purity, Soul Food, and Sunni Islam: Explorations at the Intersection of Consumption and Resistance," *Cultural Anthropology* 19, (2004).

63. Caroline Walker Bynum, *Holy Feast and Holy Fast: The Religious Significance of Food to Medieval Women* (Berkeley: University of California Press, 1987). See also her shorter piece, "Fast, Feast, and Flesh: The Religious Significance of Food to Medieval Women," in *Food and Culture: A Reader,* ed. C. Counihan and P. van Esterik (New York: Routledge, 2008).

64. A different son of Abraham, Isaac, is named as the possible sacrifice in the Old Testament.

65. For a full description of the ceremony and the foods, see *The Haggadah* (London: Soncino Press, 1975). There is a short summary in Carolyn Korsmeyer, "Bitter Herbs and Unleavened Bread," in *The Taste Culture Reader: Experiencing Food and Drink,* ed. C. Korsmeyer (Oxford: Berg, 2005).

66. Simoons, *Eat Not This Flesh,* 107. The original is in Mark Graubard, *Man's Food, Its Rhyme or Reason* (New York: Macmillan, 1943), 19–20.

67. See Harris, *Good to Eat,* chapter 8, "Small Things," 154–74, for a lively discussion of insect aversion and enjoyment.

CHAPTER 5 INDUSTRIALIZATION

1. Carolyn Wyman, *Spam: A Biography* (San Diego, CA: Harcourt Brace, 1999).

2. Keith Wrightson, *Earthly Necessities: Economic Lives in Early Modern Britain, 1470–1750* (London: Penguin Books, 2000), 3.

3. Thomas More, *Utopia,* ed. and trans. D. Wooton (Indianapolis, IN: Hackett, 1999), 66–68.

4. Adam Smith, *The Wealth of Nations,* books 1–3 (London: Penguin, 1987), 279, 269.

5. Karl Marx, *Capital* (London Penguin, 1990), 874.

6. Ibid., 874–75.

7. In England, the wool trade in the medieval and early modern period was a key precursor of the development of industrial capitalism and, as noted earlier, required the use of land not for the purposes of raising food but producing fleeces. See Peter J. Bowden, *The Wool Trade in Tudor and Stuart England* (London: Routledge, 2006).

8. Max Weber, *The Protestant Ethic and the Spirit of Capitalism,* trans. T. Parsons (London: Routledge, 1992).

9. See Hendrik Spruyt, *The Sovereign State and Its Competitors: An Analysis of Systems Change* (Princeton, NJ: Princeton University Press, 1994).

10. The Bank of England was the first central bank as such, founded in 1694.

11. Eric Hobsbawm, *The Age of Revolution, 1789–1848* (London: Weidenfeld and Nicolson, 1975), 28–29.

12. T.S. Ashton, *The Industrial Revolution* (Oxford: Oxford University Press, 1968), 21–2. J.D. Chambers and G.E. Mingay, *The Agricultural Revolution, 1750–1880* (London: B.T. Batsford, 1966), 40–41.

13. M.W. Flinn, *Origins of the Industrial Revolution* (London: Longman, 1966), 22.

14. Hans-Jurgen Teuteberg and Jean-Louis Flandrin, "Transformation of the European Diet" in *Food: A Culinary History from Antiquity to the Present,* ed. J.-L. Flandrin and M. Montanari (New York, NY: Columbia University Press, 1999), 448.

15. Ashton, *Industrial Revolution,* 21; Chambers and Mingay, *Agricultural Revolution,* 72.

16. Some historians question whether changes to agricultural were the major dynamic force in providing the conditions for the Industrial Revolution, preferring instead to focus on changes in industry, urbanization, and commercialization. See Pat Hudson, *The Industrial Revolution* (London: Edward Arnold, 1992), 97; Robert Allen, *The British Industrial Revolution in Global Perspective* (Cambridge: Cambridge University Press, 2009), 78–79.

17. See Michael Zmolek, *Rethinking the Industrial Revolution: Five Centuries of Transition from Agrarian to Industrial Capitalism in England* (Chicago: Haymarket, 2014).

18. Flinn, *Origins of the Industrial Revolution,* 96–97.

19. Wrightson, *Earthly Necessities,* 29.

20. Maxine Berg, *The Age of Manufactures: Industry, Innovation, and Work in Britain, 1700–1820* (London: Fontana, 1985).

21. Ashton, *Industrial Revolution,* 2.

22. Eric Pawson, *The Early Industrial Revolution: Britain in the Eighteenth Century* (London: Batson, 1979), 193.

23. Ibid., 199.

24. See Walter Shelton, *English Hunger and Industrial Disorders: A Study of Social Conflict During the First Decade of George III's Reign* (London: Macmillan, 1973). In a famous essay, E.P. Thompson argued that the food riots of the eighteenth century represented the resistance of workers to commercializing practices in the defence of traditional rights and customs. There was a "moral economy" surrounding food production and distribution that was being affronted by the rise of capitalist farming. E.P. Thompson, "The Moral Economy of the English Crowd in the Eighteenth Century," *Past & Present* 50 (1971): 76–136.

25. Jeffrey M. Pilcher, *Food in World History* (London: Routledge, 2006), 56.

26. Reay Tannahill, *Food in History* (London: Penguin, 1988), 291–95.

27. As Ruth Schwartz Cohan writes, the terms *housewife* and *husband* originate in England in the thirteenth century. A husband is the spouse of a housewife, being bonded to the *hus*, "house." These were terms applied not to the households of aristocrats, but to those of small farmers—both husband and wife were tied to and worked on the land. Industrial capitalism transforms the household into a sphere of consumption and shifts the sphere of production to its outside, turning the husband into a wage earner and the house and housewife into the mere means of his reproduction. Ruth Schwarz Cohan, *More Work for Mother: The Ironies of Household Technology from the Open Hearth to the Microwave* (London: Free Association, 1989), 17.

28. See Peter Borsay, *The English Urban Renaissance: Culture and Society in the Provincial Town, 1660–1770* (Oxford: Clarendon, 1989).

29. See Nicholas Faith, *The World the Railways Made* (London: Bodley Head, 1990); Douglas Burgess, *Engines of Empire: Steamships and the Victorian Imagination* (Stanford, CA: Stanford University Press, 2016).

30. Tom Standage, *The Victorian Internet: The Remarkable Story of the Telegraph and the Nineteenth Century Online Pioneers* (London: Weidenfeld and Nicolson, 1998).

31. See Elliott Young, *Alien Nation: Chinese Migration in the Americas from the Coolie Era Through World War II* (Chapel Hill: University of North Carolina Press, 2014).

32. See, for example, Donna R. Gabaccia, *We Are What We Eat: Ethnic Food and the Making of Americans* (Cambridge, MA: Harvard University Press, 1998); and Panikos Panayi, *Spicing Up Britain: The Multicultural History of British Food* (London: Reaktion, 2008).

33. William Cronon, *Nature's Metropolis* (New York: W. W. Norton, 1991), 70.

34. Ruth Schwarz Cohan, *A Social History of American Technology* (Oxford: Oxford University Press, 1997), 105–12.

35. Cronon, *Nature's Metropolis*, 103.

36. Alan Marcus and Howard Segal, *Technology in America: A Brief History* (San Diego, CA: Harcourt, 1989), 189–90.

37. Deborah Fitzgerald, "Farmers Deskilled: Hybrid Corn and Farmers' Work" in *Technology and American History,* ed. S. H. Cutcliffe and T. S. Reynolds (Chicago: University of Chicago Press, 1997).

38. Cronon, *Nature's Metropolis*, 210.

39. Upton Sinclair, *The Jungle* (London: Penguin, 2002).

40. Cronon, *Nature's Metropolis*, 111.

41. Jürgen Osterhammel, *The Transformation of the World: A Global History of the Nineteenth Century* (Princeton, NJ: Princeton University Press, 2014), 277.

42. See David Hounshell, *From the American System to Mass Production: The Development of Manufacturing Technology in the United States* (Baltimore, MD: Johns Hopkins University Press, 1984), chapter 6. For the wider cultural effects of the automobile, see Rudi Volti, *Cars and Culture: The Life Story of a Technology* (Baltimore, MD: Johns Hopkins University Press, 2006).

43. Giorgio Pedrocco, "The Food Industry and New Preservation Techniques," in *Food: A Culinary History from Antiquity to the Present*, ed. J.-L. Flandrin and M. Montanari (New York: Columbia University Press, 1999), 486; Maguelonne Toussaint-Samat, *A History of Food*, trans. A. Bell (Oxford: Wiley-Blackwell, 2009), 664–66.

44. Toussaint-Samat, *A History of Food*, 667; Tannahill, *Food in History*, 310.

45. Tannahill, *Food in History*, 311.

46. Tim Buxbaum, *Icehouses* (London: Shire Library, 2014); Tom Jackson, *Chilled: How Refrigeration Changed the World and Might Do So Again* (London: Bloomsbury Sigma, 2015).

47. For a more detailed description of the process, see Jackson, *Chilled*, 168–71.

48. Jonathan Rees, *Refrigeration Nation: A History of Ice, Appliances, and Enterprise in America* (Baltimore, MD: Johns Hopkins University Press, 2013).

49. Tannahill, *Food in History*, 319; Andrew Smith, *Sugar: A Global History* (London: Reaktion Books, 2015).

50. Tannahill, *Food in History*, 317–18; J. H. van Stuyvenberg, *Margarine: An Economic, Social, and Scientific History, 1869–1969* (Liverpool: University of Liverpool Press, 1969).

51. See Bernard Doray, *From Taylorism to Fordism: A Rational Madness?* trans. D. Macey (London: Free Association, 1988).

52. For two very different perspectives on the relationship between nationalism, modernization, and industrialization, see Ernest Gellner, *Nations and Nationalism* (Oxford: Blackwell, 2006), and Michel Foucault, *Security, Territory, Population: Lectures at the College de France 1977–1978*, trans G. Burchell (London: Palgrave Macmillan, 2007).

53. See Matthew Parker, *The Sugar Barons: Family, Corruption, Empire and War in the West Indies* (New York: Walker, 2011); Bartow Elmore, *Citizen Coke: The Making of Coca-Cola Capitalism* (New York: Norton, 2015)

54. Tannahill, *Food in History*, 328–31; see also Katherine Parkin, *Food Is Love: Advertising and Gender Roles in Modern America* (Philadelphia: University of Pennsylvania Press, 2006).

55. For more details of the process involved, see Koert van Mensvoort and Hendrik-Jan Grievink, *The In Vitro Meat Cookbook* (Amsterdam: BIS Publishers, 2014).

56. Henry Fountain, "Building a $325,000 Burger," *New York Times*, May 12, 2013: www.nytimes.com/2013/05/14/science/engineering-the-325000-in-vitro-burger.html; and Alok Jha, "Synthetic Meat: How the World's Costliest Burger Made It on to the Plate," *Guardian*, August 5, 2013, www.theguardian.com/science/2013/aug/05/synthetic-meat-burger-stem-cells.

57. See Neil Stephens, "Growing Meat in the Laboratory: The Promise, Ontology, and Ethical Boundary-Work of Using Muscle Cells to Make Food," *Configurations* 21, (2013): 159–81.

58. See Guy Cook, *Genetically Modified Language: The Discourse of Arguments for GM Crops and Food* (London: Routledge, 2004).

59. Klaus Schwab, *The Fourth Industrial Revolution* (Geneva: World Economic Forum, 2016); for a more sceptical view of these processes, see Martin Ford, *The Lights in the Tunnel: Automation, Accelerating Technology and the Economy of the Future* (N.P.: Acculant Publishing, 2009).

CHAPTER 6 THE PUBLIC SPHERE

1. Samuel Pepys, *The Diaries of Samuel Pepys: A Selection,* ed. R. Latham (London: Penguin, 2003), 5.

2. See Claire Tomalin, *Samuel Pepys: The Unequalled Self* (London: Penguin, 2003).

3. N.D. Fustel de Coulanges, *The Ancient City: A Study of the Religion, Laws, and Institutions of Greece and Rome* (Baltimore, MD: Johns Hopkins University Press, 1980), 147–48; Aristotle, *The Politics and the Constitution of Athens,* trans. B. Jowett (Cambridge: Cambridge University Press, 1996), 1271a, 28–38.

4. R.E. Wycherley, *How the Greeks Built Cities* (London: Macmillan, 1962), 65–66.

5. Jérôme Carcopino, *Daily Life in Ancient Rome: The People and the City at the Height of Empire* (Harmondsworth: Penguin, 1970), 37–38.

6. Norbert Elias, *The Civilizing Process,* trans. E. Jephcott (Oxford: Blackwell, 2000); Jürgen Habermas, *The Structural Transformation of the Public Sphere,* trans. T. Burger (Cambridge: Polity, 1992).

7. Ibid., 21.

8. Ibid., 30.

9. Ibid., 33.

10. Hannah Arendt, *The Human Condition* (Chicago: University of Chicago Press, 1998), 213.

11. Richard Sennett, *The Fall of Public Man* (London: Penguin, 2002), 81, 82.

12. Arendt, *The Human Condition,* 126–35.

13. Beat Kümin, *Drinking Matters: Public Houses and Social Exchange in Early Modern Central Europe* (London: Palgrave Macmillan, 2007), 2.

14. Mark Hailwood, *Alehouses and Good Fellowship in Early Modern England* (Woodbridge, U.K.: Boydell, 2014), 2–3.

15. Ibid., 3.

16. Peter Clarke, *The English Alehouse: A Social History* (London: Longman, 1983), 3.

17. Hailwood, *Alehouses,* 4.

18. Some estimates show that in the fourteenth and fifteenth centuries, English people consumed eight pints of ale a day, on average, while by the end of the seventeenth, it had fallen to just two. Hailwood, *Alehouses,* 4–5.

19. Ibid., 5–6.

20. See Angela McShane, "Material Culture and 'Political Drinking' in Seventeenth-Century England," *Past and Present* Supplement 9 (2014): 247–76; James Nicholls, *The Politics of Alcohol: A History of the Drink Question in England* (Manchester, UK: Manchester University Press, 2011), chapter 2.

21. Nicholls, *Politics of Alcohol,* 28–29.

22. Hailwood, *Alehouses,* 1.

23. James Brown, "Ale House Licensing and State Formation in Early Modern England," in *Intoxication and Society: Problematic Pleasures of Drugs and Alcohol,* ed. J. Herring, C. Regan, D. Weinberg, and P. Withington, 110–32 (London: Palgrave Macmillan, 2013).

24. Peter Haydon, *The English Pub: A History* (London: Robert Hale, 1994), 104.

25. Daniel Vasey, *The Pub and English Social Change* (New York: AMS, 1990), 37.

26. Ruth Cherrington, *Not Just Beer and Bingo: A Social History of Workingmen's Clubs* (Bloomington, IN: Author House, 2012).

27. Coffee began to be used as a beverage in Arabia in the mid-fifteenth century via Ethiopia, one (if not the sole) site of the coffee plant's first cultivation. See Ralph Hattox, *Coffee and Coffeehouses: The Origins of a Social Beverage in the Medieval Near East* (Seattle: University of Washington Press, 1985).

28. Quoted in Markman Ellis, *The Coffee-House: A Cultural History* (London: Weidenfeld & Nicolson, 2004).

29. Brian Cowan, "*Café* or Coffee-House? Transnational Histories of Coffee and Sociability," in *Drinking in the Eighteenth and Nineteenth Centuries,* ed. S. Schmid and B. Schmidt-Haberkamp, 35–46 (London: Pickering and Chatto, 2014).

30. Brian Cowan, *The Social Life of Coffee: The Rise of the Emergence of the English Coffee-House* (New Haven, CT: Yale University Press, 2005), 12–13.

31. Perry Anderson, "Origins of the Present Crisis," *New Left Review* 1, (1964): 26–53.

32. Sennett, *Fall of Public Man,* 215.

33. Ibid., 83.

34. Ellis, *The Coffee-House,* 174.

35. Cowan, *The Social Life of Coffee,* 165.

36. Ibid., 169.

37. Ellis, *The Coffee-House,* 183.

38. See Antony Wild, *Black Gold: The Dark History of Coffee* (London: Harper Perennial, 2005).

39. Frank Trentmann, *Empire of Things: How We Became a World of Consumers, from the Fifteenth Century to the Twenty-First* (London: Allen Lane, 2016), 92.

40. See Reinhart Koselleck, *Critique and Crisis: Enlightenment and the Pathogenesis of Modern Society* (Cambridge, MA: MIT Press, 1988); James van Horn Melton, *The Rise of the Public in Enlightenment Europe* (Cambridge: Cambridge University Press), 224; Cowan, *The Social Life of Coffee,* 256.

41. W. Scott Haine, *The World of the Paris Café: Sociability among the French Working Class, 1789–1914* (Baltimore, MD: The Johns Hopkins University Press, 1996); Charlotte Ashby, Tag Gronberg, and Simon Shaw-Miller, eds., *The Viennese Café and Fin-de-Siècle Culture* (New York: Berghahn, 2013).

42. Rebecca Spang, *The Invention of the Restaurant: Paris and Modern Gastronomic Culture* (Cambridge, MA: Harvard University Press, 2000).

43. The restaurants of nineteenth-century Paris maintained spaces for private dining away from the theater of the dining room in the form of the *cabinets*

particuliers. These private rooms were often depicted as sites of debauchery and political intrigue, but for Spang, they mark the transformation of the restaurant from a semi-public to a semi-private space as it became increasingly depoliticised after the Revolution. Spang, *Invention,* 232–33.

44. Ibid., 33.

45. See Priscilla Ferguson, *Accounting for Taste: The Triumph of French Cuisine* (Chicago: Chicago University Press, 2004).

46. Rachel Rich, *Bourgeois Consumption: Food, Space, and Identity in London and Paris, 1850–1914* (Manchester, U.K.: Manchester University Press, 2011).

47. Trentmann, *Empire,* 13–14.

48. "Behind the Scenes: The World's First Michelin-Starred Street Food Stall," *Michelin Guide,* October 4, 2016, https://guide.michelin.com.hk/en/behind-the-scenes-the-world-s-first-michelin-starred-street-food-stall.

49. Beat Kümin, "Eating Out before the Restaurant: Dining Cultures in Early-Modern Inns," in *Eating Out in Europe: Picnics, Gourmet Dining, and Snacks Since the Late Eighteenth Century,* ed. M. Jacobs and P. Scholliers, 71–88 (Oxford: Berg, 2003).

50. See Robert Applebaum, *Dishing It Out: In Search of the Restaurant Experience* (London: Reaktion Books, 2011).

51. See Stephen Mennell, *All Manners of Food: Eating and Taste in England and France from the Middle Ages to the Present,* 2nd ed. (Urbana: University of Illinois Press, 1996); Pierre Bourdieu, *Distinction: A Social Critique of the Judgement of Taste,* trans. R. Nice (Cambridge, MA: Harvard University Press, 1984).

52. Bourdieu, *Distinction.*

53. David Bell and Gill Valentine, *Consuming Geographies: We Are Where We Eat* (London: Routledge, 1997).

54. See Ann Murcott, "Family Meals: A Thing of the Past?" in *Food, Health, and Identity,* ed. P. Caplan, 32–49 (London: Routledge, 1997).

55. Adel den Hartog, "Technological Innovations and Eating Out As a Mass Phenomenon in Europe: A Preamble," in *Eating Out in Europe: Picnics, Gourmet Dining, and Snacks Since the Late Eighteenth Century,* ed. M. Jacobs and M. Scholliers, 263–80 (Oxford: Berg, 2003).

56. See, for example, Robert Putnam, *Bowling Alone: The Collapse and Revival of American Community* (New York: Simon & Schuster, 2000).

57. See David Harvey, *Spaces of Global Capitalism: Towards a Theory of Uneven Geographic Development* (London: Verso, 2006).

58. Arjun Appadurai, introduction to *The Social Life of Things: Commodities in Cultural Perspective,* ed. A. Appadurai (Cambridge: Cambridge University Press, 1986).

CHAPTER 7 THE MODERN STATE

1. Gin distilled in London was said to have its origins in the Netherlands, where the spirit *genever* became a favorite of soldiers fighting the Spanish in the Dutch Revolt in the late sixteenth century. Lesley Jacobs Solomonson, *Gin: A Global History* (London: Reaktion Books, 2012), 23.

2. While it would be an anachronism to describe him as a "liberal," the classic account of this kind of state—or sovereign power—was given by Thomas Hobbes in 1651 in his *Leviathan* (Cambridge: Cambridge University Press, 1991).

3. Michel Foucault, *The Birth of Biopolitics: Lectures at the Collège de France, 1978–9*, trans. G. Burchell (Basingstoke, U.K.: Palgrave Macmillan, 2008).

4. A great deal of beer consumed in the Middle Ages was "small beer," containing very low amounts of alcohol. This could be drunk at volume and was deemed more suitable for children and servants.

5. Ethanol cannot be distilled to 100 percent ABV, because at around 95 percent ABV it forms an azeotrope with water; the proportions of water and ethanol in the vapour released by boiling remain constant after condensation.

6. Though animal intoxication may be rarer than sometimes thought. There is also the question of whether any behavioral effects of alcohol on animals can be counted as "drunkenness" without anthropomorphization. See Jason G. Goldman, "Do Animals Like Drugs and Alcohol?," *BBC*, May 28, 2014, *http://www.bbc.com/future/story/20140528-do-animals-take-drugs*; Rod Phillips, *Alcohol: A History* (Chapel Hill: University of North Carolina Press, 2014), 7.

7. Phillips, *Alcohol*, 8; Patrick McGovern, *Ancient Wine: The Search for the Origins of Viniculture* (Princeton, NJ: Princeton University Press, 2003), 10–11. Evidence from Mexico suggests that the earliest alcoholic beverages may have been made not by fermenting fruit, but by the mastication and expectoration of carbohydrate-rich foods, such as wild grains, tubers, and grasses. Human saliva creates enzymes that work like yeast on carbohydrates, breaking them down into alcohol and carbon dioxide. In the Amazon rainforest, the chewing and spitting out of cassava or yuca is still used to start off the fermentation process. See Patrick McGovern, review of *Alcohol: A History,* by Rod Phillips, *Journal of Interdisciplinary History* 46, (2015): 105–107, 106; and Mollie Bloudoff-Indelicato, "Ancient Alcoholic Drink's Unusual Starter: Human Spit," *National Geographic,* March 30, 2015, http://theplate.nationalgeographic.com/2015/03/30/ancient-alcoholic-drinks-unusual-starter-human-spit.

8. An exception being the late medieval and early modern period, in which distilling and brewing was carried out at a significant scale. Phillips, *Alcohol*, 101.

9. A. Lynn Martin, *Alcohol, Violence, and Disorder in Traditional Europe* (Kirksville, MO: Truman University State Press, 2009).

10. Max Harris, *Carnival and Other Christian Festivals: Folk Theology and Folk Performance* (Austin: University of Texas Press, 2003).

11. Sami Zubaida, "Drink, Meals and Social Boundaries," in *Food Consumption in Global Perspective: Essays in the Anthropology of Food in Honour of Jack Goody,* eds. Jakob Klein and Anne Murcott, 209–23 (London: Palgrave, 2014), 211–12.

12. Rudi Matthee, "Alcohol in the Islamic Middle East: Ambivalence and Ambiguity," in *Cultures of Intoxication,* ed. P. Withington and A. McShane, 100–126 (Oxford: Oxford University Press, 2014), 111–12.

13. Tannahill, *Food in History*, 243–44.

14. Phillips, *Alcohol: A History*, 110–11.

15. Jan-Willem Gerritsen, *The Control of Fuddle and Flash: A Sociological History of the Regulation of Alcohol and Opiates* (Leiden: Brill, 2000), 30–31.

16. Hendrik Spruyt, *The Sovereign State and Its Competitors* (Princeton, NJ: Princeton University Press, 1994); Max Weber, "Politics As a Vocation," in *From Max Weber*, ed. H.H. Gerth and C. Wright Mills, 77–128 (London: Routledge, 1991).

17. Control of tax revenues was, of course, a prerequisite of funding effective armed forces, an ever-costlier concern as military technologies and strategies led to the prolongation and complication of warfare. See William McNeill, *The Pursuit of Power: Technology, Armed Force, and Society Since 1000 A.D.* (Chicago: University of Chicago Press, 1984); Brian Downing, *The Military Revolution and Political Change: The Origins of Democracy and Autocracy in Early Modern Europe* (Princeton, NJ: Princeton University Press, 1992); Charles Tilly, *Coercion, Capital, and European States, A.D. 990–1992* (Oxford: Wiley-Blackwell, 1990); and Janice E. Thompson, *Mercenaries, Pirates, and Sovereigns: State-Building and Extra-Territorial Violence in Early Modern Europe* (Princeton, NJ: Princeton University Press, 1994).

18. Patrick Dillon, *The Much Lamented Death of Madam Geneva: The Eighteenth Century Gin Craze* (London: Review, 2002), 9–10.

19. Peter Clark, "The Mother Gin Controversy in the Early Eighteenth Century," *Transactions of the Royal Historical Society* 38 (1988): 63–84, 68.

20. Jessica Warner, *Craze: Gin and Debauchery in an Age of Reason* (London: Profile, 2003), 17, 29.

21. In 1729, around five million gallons of spirits were being drunk in England every year, and it is possible that between the 1720s and 1740s, on average, individual Londoners (men, women, and children) were drinking between one and two pints of gin a week. James Nicholls, *The Politics of Alcohol: A History of the Drink Question in England* (Manchester, U.K.: Manchester University Press, 2009), 36.

22. M. Dorothy George, *London Life in the Eighteenth Century* (London: Penguin, 1992), 42; T.S. Ashton, *An Economic History of England: The Eighteenth Century* (London: Methuen, 1955), 6.

23. Clark, "The Mother Gin Controversy," 71.

24. Stanley Cohen's famous concept of a moral panic illuminates much about the gin craze, particularly given the limited evidence of a real rise in drunkenness and drink-related crime in the period. As sociologists like Stuart Hall use the term, a moral panic is a *perception* of the emergence of a new kind of deviant activity that often expresses an ideological response to underlying social and economic crises. See Stanley Cohen, *Folk Devils and Moral Panics: The Creation of Mods and Rockers* (London: Routledge, 2002); and Stuart Hall, Chas Critcher, Tony Jefferson, John Clarke, and Brian Roberts, *Policing the Crisis: Mugging, the State, and Law and Order* (London: Palgrave Macmillan, 2013).

25. Warner, *Craze*, 96–97, 105.

26. Dillon, *Madam Geneva*, chapter 3.

27. Warner, *Craze*, chapters 6, 7; Dillon, *Madam Geneva*, 220.

28. Warner, *Craze*, 224.

29. Hogarth's central figure in *Gin Lane* may well have alluded to the infamous case of Judith Defour, accused and convicted in 1734 of murdering her baby daughter, whom she had taken out from the workhouse for the day, in order to sell her clothes to pay for gin. Dillon, *Madam Geneva*, 93–97, 208.

30. Adam Smith was clear that the purpose of the mode of inquiry he called "political economy" was not simply to demonstrate how people could provide for themselves through the market, but, as a "science of a statesman or legislator," it was also concerned with how "to supply the state or commonwealth with a revenue sufficient for the public services. It proposes to enrich both the people and the sovereign." Adam Smith, *The Wealth of Nations, Books IV-V* (London: Penguin, 1999), 5.

31. Dillon, *Madam Geneva*, 275–76.

32. Nicholls, *The Politics of Alcohol*, 39–40.

33. Warner, *Craze*, 90–91.

34. See Michel Foucault, "Governmentality," in *The Foucault Effect: Studies in Governmentality,* ed. G. Burchell, C. Gordon, and P. Miller (Chicago: University of Chicago Press, 1991).

35. Michel Foucault, *Society Must Be Defended: Lectures at the Collège de France, 1975-6,* trans. D. Macey (London: Penguin, 2004), 35.

36. Foucault, *Society*, 35.

37. Michel Foucault, *Discipline and Punish,* trans A. Sheridan (London: Penguin, 1991), 138.

38. Foucault, *Discipline and Punish,* 200–208.

39. Foucault, *The Birth of Biopolitics,* 317.

40. Daniel Okrent, *Last Call: The Rise and Fall of Prohibition* (New York: Scribner, 2010), 111–12.

41. W.J. Rorabaugh, *The Alcoholic Republic: An American Tradition* (Oxford: Oxford University Press, 1979), 9–11.

42. Sharon Salinger, *Taverns and Drinking in Early America* (Baltimore, MD: Johns Hopkins University Press, 2002), 4–5.

43. Alexis de Tocqueville, *Democracy in America* (London: Penguin, [1835; 1840] 2003), 49–51.

44. Salinger, *Taverns*, 5.

45. K. Austin Kerr, *Organized for Prohibition: A New History of the Anti-Saloon League* (New Haven, CT: Yale University Press, 1985).

46. Earlier temperance organizations, such as the Woman's Christian Temperance Union and the Prohibition Party, had adopted a broad range of arguments on moral and political questions, with the Prohibition Party fielding candidates in elections. Okrent, *Last Call*, 36.

47. Okrent, *Last Call*, 46–52; Kerr, *Organized for Prohibition*, 7–11.

48. Richard Hofstadter, *The Age of Reform: From Bryan to FDR* (London: Cape, 1962); Andrew Sinclair, *Prohibition: The Era of Excess* (London: Faber and Faber, 1962).

49. Joseph Gusfield, *Symbolic Crusade: Status Politics and the American Temperance Movement* (Urbana: University of Illinois Press, 1963).

50. Norman Clarke, *Deliver Us from Evil: An Interpretation of American Prohibition* (New York: Norton, 1976), 12–13.

51. Kerr, *Organization,* 279.

52. Prohibition had the opposite effect on rates of crime than its proponents had foreseen, not only leading to the open flouting of the law in many places, but also placing considerable political and economic muscle in the hands of the mafia. Thomas Reppetto, *American Mafia: A History of Its Rise to Power* (New York: Henry Holt, 2004).

53. Lisa McGirr, *The War on Alcohol: Prohibition and the Rise of the American State* (New York, NY: Norton, 2016).

54. Maria Valverde, *Diseases of the Will: Alcohol and the Dilemmas of Freedom* (Cambridge: Cambridge University Press, 1988), 2.

55. Jean-Charles Sournia, *A History of Alcoholism* (Oxford: Basil Blackwell, 1990), 48–49.

56. Richard Davenport-Hines, *The Pursuit of Oblivion: A Social History of Drugs* (London: Phoenix, 2002), 237.

57. The self-government of addiction is most clearly articulated in the two key notions of Alcoholics Anonymous: self-control and recovery. Valverde, *Diseases of the Will,* 125–26.

CHAPTER 8 IDENTITY

1. Ernest Gellner, *Nations and Nationalism* (Blackwell: Oxford, 1984) and Eric J. Hobsbawm, *Nations and Nationalism Since 1780* (Cambridge: Cambridge University Press, 1990).

2. See Hobsbawm, *Nations and Nationalism,* 18–22; and Graham Robb, *The Discovery of France* (London: Picador, 2007).

3. Gellner, *Nations and Nationalism,* 39–52.

4. Eric J. Hobsbawm and Terence Ranger, *The Invention of Tradition* (Cambridge: Cambridge University Press, 1992).

5. Sami Zubaida, "Hazz al-Quhuf: An Urban Satire on Peasant Life in Seventeenth-Century Egypt," in *Food between the Country and the City: Ethnographies of a Changing Global Foodscape,* ed. Nuno Domingo, Jose Manuel Sorbal, and Harry West, 161–74 (London: Bloomsbury, 2014).

6. Together with Eric Hobsbwam and Ernest Gellner, Benedict Anderson's classic study of the construction of the nation as an "imagined community" through print media, novels, museums, school textbooks, maps, and the census is an exemplar of this approach to the subject—*Imagined Communities: Reflections on the Origin and Spread of Nationalism* (London: Verso, 1991).

7. A good summary can be found in Atsuko Ichijo and Ronald Ranta, *Food, National Identity and Nationalism: From Everyday to Global Politics* (Basingstoke, U.K.: Palgrave Macmillan, 2016).

8. John Dickie, *Delizia: The Epic History of Italians and Their Food* (New York: Free Press, 2008), 196–215.

9. Dickie, *Delizia,* 197.

10. Dickie, *Delizia,* 209.

11. Wayne Northcutt, "José Bové vs. McDonald's: The Making of a National Hero in the French Anti-Globalisation Movement," *Journal of the Western Society for French History* 31, 2003, https://quod.lib.umich.edu/w/wsfh/06422 92.0031.020?view=text;rgn=main.

12. Ibid.

13. Carlo Petrini, *Slow Food: The Case for Taste* (New York: Columbia University Press, 2003).

14. Sami Zubaida, "National, Communal and Global Dimensions in Middle Eastern Food Cultures," in *A Taste of Thyme: Culinary Cultures of the Middle East,* ed. Sami Zubaida and Richard Tapper (London: Tauris, 2000), 44–45.

15. Dickie, *Delizia,* 216–19.

16. Dickie, *Delizia,* 224–26.

17. Charles Perry, "The Taste for Layered Bread among the Nomadic Turks and the Central Asian Origins of Baklava," in Zubaida and Tapper, *A Taste of Thyme,* 87–92.

18. Perry, "The Taste for Layered Bread."

19. See Florence Ollivry, *Les secrets d'Alep: Une grande ville arabe révélée par sa cuisine* (Paris: Sinbad/Actes Sud, 2006). While it concentrates on Aleppo, it is clear about its regional kinship to southern Anatolia and northern Iraq, as well as its ethnic mix. Most sources relate to particular "national" and ethnic cuisines and particular regions within countries, in addition to the "Mediterranean."

20. Ollivry, *Les secrets d'Alep,* 34–37.

21. Bruce Masters, "Aleppo: The Ottoman Empire's Caravan City," in *The Ottoman City between East and West,* ed. Edhem Eldem, Daniel Goffman, and Bruce Masters, (Cambridge: Cambridge University Press 1999), 32–34.

22. Among the many examples, see, for instance, Deniz Gursoy, *Turkish Cuisine in Historical Perspective,* trans. Joyce Matthews (Istanbul: Ogla, 2006); Ghillie Basan, *Classic Turkish Cookery* (London: Tauris, 1997); and Ersu Pekin and Ayse Sumer, eds. *Timeless Tastes: Turkish Culinary Culture* (Istanbul: Divan, 2003)—the essay in this latter book by Tugrul Savkay, "The Cultural and Historic Context of Turkish Cuisine" (72–89) is particularly pertinent to the argument here.

23. Mai Yamani, "You Are What You Cook: Cuisine and Class in Mecca," in Zubaida and Tapper, *A Taste of Thyme.*

24. Lane, 1898, 159–65.

25. Liora Gvion, *Beyond Hummus and Falafel: Social and Political Aspects of Palestinian Food in Israel* (Berkeley: University of California Press, 2012).

26. Sami Zubaida, "The Idea of 'Indian Food': Between the Colonial and the Global," *Food and History* 7 (2009): 191–210.

27. Lizzie Collingham, *Curry: Tales of Cooks and Conquerors* (Oxford: Oxford University Press, 2006), 47–80; William Darlymple, *White Mughals* (London: Harper-Collins, 2003); M. N. Pearson, *The Portuguese in India* (Cambridge: Cambridge University Press, 1987).

28. Collingham, *Curry.*

29. Sunil Khilnani, *The Idea of India* (London: Penguin, 1997).

30. Arjun Appadurai, "How to Make a National Cuisine: Cookbooks in Contemporary India," *Comparative Studies in Society and History* 30 (1998): 3–24.

31. Accounts of the introduction and spread of Indian food in Britain are related, among others, by Shrabani Basu, *Curry: The Story Of Britain's Favourite Dish* (New Delhi: Harper Collins, 1999); and Collingham, *Curry*.

32. Basu, *Curry*, 118–19; and Collingham, *Curry*, 154.

33. This account draws on personal observation of the British food and restaurant scene from the later decades of the twentieth century to the present by Sami Zubaida.

34. Eugene Weber, *Peasants into Frenchmen* (London: Chatto, 1997); and Dickie, *Delizia*.

35. "Mediterranean Diet," *UNESCO*, www.unesco.org/culture/ich/en/RL/mediterranean-diet-00884.

36. Fernand Braudel, *The Mediterranean and the Mediterranean World in the Age of Philip II*, trans. Sian Reynolds, 2 vols. (London: Collins 1973).

37. Suraiya Faroqhi, *Towns and Townsmen of Ottoman Anatolia: Trades, Crafts and Food Production in Urban Settings, 1520–1650* (Cambridge: Cambridge University Press, 1984), 215.

CHAPTER 9 DISTINCTION

1. Pierre Bourdieu, *Distinction: A Social Critique of the Judgement of Taste*, trans. Richard Nice (Cambridge, MA: Harvard University Press, 1984).

2. Jack Goody, *Cooking, Cuisine and Class: A Study in Comparative Sociology* (Cambridge: Cambridge University Press, 1982).

3. Mireille Corbier, "The Broad Bean and the Moray: Social Hierarchies and Food in Rome," in *Food: A Culinary History From Antiquity to the Present*, ed. J.-L. Flandrin and M. Montanari, 128–40 (New York: Columbia University Press, 1999).

4. Ibid., 129–30.

5. Antoni Riera-Melis, "Society, Food and Feudalism," in *Food: A Culinary Culinary History from Antiquity to the Present*, ed. J.-L. Flandrin and M. Montanari, 251–67 (New York: Columbia University Press, 1999), 262–63.

6. Norbert Elias, *The Civilizing Process: The History of Manners and State Formation and Civilization* (Oxford: Blackwell, 1994). See also the exposition of Elias's theory in Stephen Mennell, *All Manners of Food: Eating and Taste in England and France from the Middle Ages to the Present* (Oxford: Blackwell, 1985), 20–39.

7. Elias, *Civilizing Process*, 45.

8. Ibid., 51–52.

9. Massimo Montanari, "Introduction: Toward a New Dietary Balance," in *Food: A Culinary History from Antiquity to the Present*, ed. J.-L. Flandrin and M. Montanari, 247–50 (New York: Columbia University Press, 1999), 249.

10. John Dickie, *Delizia: The Epic History of Italians and Their Food* (New York: Free Press, 2008), 39–56.

11. See Rachel Laudan, *Cuisine and Empire: Cooking in World History* (Berkeley: University of California Press, 2013) 175–77.

12. See, for instance, Eugen Weber, *Peasants into Frenchmen: The Modernisation of Rural France: 1870–1914* (London: Chatto, 1977).

13. Dickie, *Delizia*, 66–67; 70–71.

14. Ibid., 87.

15. Ibid., 130.

16. The succession of cookery manuals by the prominent cooks over those centuries and their relation to the kitchens of royalty and notability is elucidated in Barbara Ketcham Wheaton, *Savoring the Past: The French Kitchen and Table from 1300 to 1789*. (London: Chatto and Windus, 1983).

17. Ibid., 113, 137–39.

18. Ibid., 132–38.

19. Ibid., 137.

20. Stephan Yerasimos, *A la table du Grand Turc* (Paris: Sinbad, 2001).

21. Ibid., 13–16.

22. Ibid., 24–33.

23. Ibid., 25; see also Sami Zubaida, "Drink, Meals and Social Boundaries," in *Food Consumption in Global Perspective: Essays in the Anthropology of Food in Honour of Jack Goody,* ed. J.A. Kleinand and A. Murcott, 209–23. (London: Palgrave, 2014).

24. Yerasimos, *A la table,* 20–21.

25. Ibid., 33–38.

26. Ibid., 34–43.

27. Wheaton, *Savouring the Past,* 160–72; Laudan, *Cuisine and Empire,* 219–22.

28. Wheaton, *Savouring the Past,* 163–64.

29. Mennell, *All Manners of Food,* 157–65.

30. Wheaton, *Savouring the Past,* 224–26.

31. Ibid., 121–26.

32. Pierre Bourdieu, *Distinction: A Social Critique of the Judgement of Taste,* trans. Richard Nice (Cambridge, MA: Harvard University Press 1984).

33. Ibid., 175–93.

34. Josee Johnston and Shyon Baumann, *Foodies: Democracy and Distinction in the Gourmet Foodscapes* (London: Routledge 2010).

35. Ibid., 3.

CHAPTER 10 POLITICAL ECONOMY

1. Geoff Tansey and Tony Worsley, *The Food System: A Guide* (London: Earthscan, 1996). See also Geoff Tansey's excellent companion website: www. tansey.org.uk.

2. A good overview of this concept and the debates it has brought in its wake can be found in André Magnan, "Food Regimes," in *The Oxford Handbook of Food History,* ed. Jeffrey M. Pilcher, 370–88 (Oxford: Oxford University Press, 2012).

3. Harriet Friedmann and Philip McMichael, "Agriculture and the State System: The Rise and Fall of National Agricultures, 1870 to the Present," *Sociologia Ruralis* 29, (1989): 95. Michel Agelietta's *A Theory of Capitalist Regulation* (London: New Left Books, 1979) is an early exponent of this approach.

4. The following paragraph summarizes Harriet Friedmann's historical account in "Feeding the Empire: The Pathologies of Globalized Agriculture," in

The Empire Reloaded: The Socialist Register 2005, ed. Leo Panitch and Colin Leys, 124–43 (London: The Merlin Press, 2004).

5. Cited in Michael Carolan, *The Sociology of Food and Agriculture,* 2nd ed. (London: Routledge, 2016), 35; and Peter Dicken, *Global Shift: Mapping the Changing Contours of the World Economy* (London: Sage, 2011), 290.

6. Friedmann, "Feeding the Empire," 131.

7. For a useful overview, see Henry Bernstein, "Food Regimes and Food Regime Analysis: A Selective Survey," (paper presented the conference Land Grabbing, Conflict and Agrarian-Environmental Transformation: Perspectives from East and Southeast Asia, Chiang Mau University, Thailand, June 2015) www.iss.nl/en/research/networks/land-deal-politics-ldpi/land-grabbing-perspectives-east-and-southeast-asia.

8. Friedmann, "Feeding the Empire," 125.

9. For good overviews, see Robert L. Heilbronner, *The Worldly Philosophers: The Lives, Times, and Ideas of the Great Economic Thinkers* (London: Penguin, 1953); and Frank Stilwell, *Political Economy: The Contest of Economic Ideas* (Oxford: Oxford University Press, 2011).

10. See, for instance, Lester R. Brown, *Full Planet, Empty Plates: The New Geopolitics of Food Scarcity* (New York: W.W. Norton, 2012).

11. The most celebrated historical advocate of this approach was Friedrich List in his *The National System of Political Economy.* Contemporary expressions of this approach include doctrines of "import-substituting industrialization," "de-linking," or "food sovereignty." See, for instance, Walden Bello, *Capitalism's Last Stand? Deglobalization in the Age of Austerity* (London: Zed, 2013).

12. Adam Smith, *The Wealth of Nations* (London: Penguin, 1982 [1776]), 118.

13. David Ricardo, *Principles of Political Economy* (New York: Dover Publications, 2004 [1817]), 77.

14. See, for instance, Anne O. Krueger, "The Political Economy of the Rent-Seeking Society," *American Economic Review* 64, (1974): 291–303.

15. Friedmann, "Feeding the Empire," 125.

16. Gary Gereffi and Miguel Korzeniewicz, eds., *Commodity Chains and Global Capitalism* (Westport, CT: Greenwood Press, 1994). See also https://globalvaluechains.org—a rich source of materials on global value chains collated by a network of researchers, activists, and policymakers.

17. See Roland Meek's classic account of the four stages of social evolution (or "development") presented by the Scottish and French Enlightenment authors in his *Social Science and the Ignoble Savage* (Cambridge: Cambridge University Press, 1974).

18. Karl Polanyi, *The Great Transformation: the Political and Economic Origins of Our Time* (Boston: Beacon Press, 1944), 79.

19. Adam Smith, *Wealth of Nations,* 121.

20. Polanyi, *Great Transformation,* 147.

21. For a comprehensive and thoroughly informed overview, see Benoit Daviron and Stefano Ponte, *The Coffee Paradox: Global Markets, Commodity Trade and the Elusive Promise of Development* (London: Zed Books, 2005).

22. Robert H. Bates, *Open-Economy Politics: The Political Economy of the World Coffee Trade* (Princeton, NJ: Princeton University Press, 1997), 137.

23. Stephen D. Krasner, "Structural Causes and Regime Consequences: Regimes As Intervening Variables," in *International Regimes,* ed. Stephen D. Krasner, 1–22, Cornell Studies in Political Economy (Ithaca, NY: Cornell University Press, 1983).

24. Bates, *Open-Economy Politics,* 151.

25. Ibid., 152.

26. Sjoerd Panhuysen and Joost Pierrot, *Coffee Barometer 2014,* 16. Available at https://hivos.org/sites/default/files/coffee_barometer_2014.pdf.

27. Daviron and Ponte, *The Coffee Paradox,* 204.

28. The original founding members in 1986 were Argentina, Australia, Brazil, Canada, Chile, Colombia, Fiji, Hungary, Indonesia, Malaysia, New Zealand, the Philippines, Thailand, and Uruguay.

29. Tony Weis, *The Global Food Economy: The Battle for the Future of Farming* (London: Zed Books, 2007), 128.

30. Excellent overviews can be found in Jennifer Clapp, *Food* (Cambridge: Polity Press, 2012); and Weis, *Global Food Economy.*

31. Philip McMichael, "Global Development and the Corporate Food Regime," in *New Directions in the Sociology of Global Development,* ed. Frederick H. Buttel and Philip McMichael, 265–99 (Oxford: Elsevier Press, 2005).

32. Philip McMichael, "A Food Regime Analysis of the 'World Food Crisis'" *Agriculture and Human Values* 26 (2009): 281–95.

33. See, for instance, Amy Trauger, ed., *Food Sovereignty in International Context: Discourse, Politics and Practice of Place* (London: Routledge, 2015).

34. Clapp, *Food,* 170.

35. "Declaration of the Forum for Food Sovereignty," Nyéléni, Mali, February 2007: https://nyeleni.org/spip.php?article290.

36. Annette A. Desmarais, *La Vía Campesina: Globalization and the Power of Peasants* (London: Pluto Press, 2007).

37. "Growing a Better Future: Food Justice in a Resource-Constrained World," *Oxfam,* June 2011, www.oxfamamerica.org/publications/growing-a-better-future.

38. Harriet Friedmann, "From Colonialism to Green Capitalism: Social Movements and the Emergence of Food Regimes," in *New Directions in the Sociology of International Development,* ed. Frederick H. Buttel and Philip D. McMichael, 227–64, Research in Rural Sociology and Development, vol. 11 (Amsterdam: Elsevier, 2005).

39. Ibid., 233.

40. See Carolan, *Sociology of Food and Agriculture,* 83–102, for an up-to-date overview.

CHAPTER 11 THE SELF

1. "Dis-mois ce que tu manges, je te dirais que tu es." Jean Anthelme Brillat-Savarin, *The Philosopher in the Kitchen* (Harmondsworth: Penguin, 1970), 13 (aphorism iv).

2. Julie Guthman, *Weighing In: Obesity, Food Justice, and the Limits of Capitalism* (Berkeley: University of California Press, 2011), 21.

3. Rachel Laudan, afterword in *Food & Foodways* 19 (2011): 164.

4. See Charlotte Biltekoff, *Eating Right in America* (Durham, NC: Duke University Press, 2013); Peter Naccarato and Kathleen LeBesco, *Culinary Capital* (London: Berg, 2012); Alice Julier, "The Political Economy of Obesity: The Fat Pay All," in *Food and Culture: A Reader,* ed. C. Counihan and P. Van Esterik, 482–99 (New York: Routledge, 2008).

5. Abigail Cope Saguy, *What's Wrong with Fat?* (New York: Oxford University Press, 2013).

6. Etienne Balibar, *Identity and Difference* (London: Verso, 2013), 56.

7. Ibid.

8. Ibid., 22.

9. Ibid., 1.

10. Alan Warde, *Consumption, Food and Taste: Culinary Antinomies and Commodity Culture* (London: Sage, 1997), 176.

11. Brillat-Savarin, *Physiologie Du Goût* (Paris: A. Sautelet, 1826).

12. *The Physiology of Taste* (San Francisco: North Point Press, 1986), 51.

13. Ibid.

14. Ibid., 173–78.

15. Ibid. See, in particular, meditations 1–2 and 21–23. Daniel Sipe, "Social Gastronomy: Fourier and Brillat-Savarin," *French Cultural Studies* 20, 2009, 222–23

16. Brillat-Savarin, *Philosopher in the Kitchen,* 214.

17. On the uses of shame in public health campaigns as well as private interactions, see, for example, R. M. Puhl, J. D. Latner, K. S. O'Brien, J. Luedicke, S. Danielsdottir, and X. R. Salas, "Potential Policies and Laws to Prohibit Weight Discrimination: Public Views from 4 Countries," *The Milbank Quarterly* 93 (2015): 691–731.

18. Brillat-Savarin, *Philosopher in the Kitchen,* 208.

19. Ibid., 217.

20. Ibid., 226. In Brillat-Savarin's view, ornamental female dining companions should be plumper than the men as a woman's "beauty consists above all in roundness of form and gracefully curving lines." This ostensibly quaint notion that women are legitimate objects of the male gaze, judged and defined by men according to their bodies, is uncomfortably persistent.

21. Biltekoff, *Eating Right in America,* 112.

22. See www.oed.com/view/Entry/129578.

23. Albert Mehrabian and Morton Wiener, "Decoding of Inconsistent Communications," *Journal of Personality and Social Psychology* 6, (1967); 109–44. Albert Mehrabian and Susan R. Ferris, "Inference of Attitudes from Nonverbal Communication in Two Channels," *Journal of Consulting Psychology* 31, 1967: 248–52.

24. Cuddy received a standing ovation for her TED talk discussing this research and demonstrating the effective poses. See TEDGlobal, "Amy Cuddy: Your Body Language May Shape Who You Are," (www.ted.com/talks/amy_cuddy_your_body_language_shapes_who_you_are).

25. Examples include Shari L. Barkin, William J. Heerman, Michael D. Warren, and Christina Rennhoff, "Millennials and the World of Work: The Impact of Obesity on Health and Productivity," *Journal of Business and Psychology* 25 (2010): 239–45; Rachel Colls and Bethan Evans, "Challenging Assumptions: Re-Thinking 'the Obesity Problem,'" *Geography* 95 (2010): 99–105; Bethan Evans, "'Gluttony or Sloth': Critical Geographies of Bodies and Morality in (Anti)Obesity Policy," *Area* 38, (2006): 259–67; Luann Heinen and Helen Darling, "Addressing Obesity in the Workplace: The Role of Employers," *The Milbank Quarterly* 87 (2009): 101–22; S.J. Robroek, K.G. Reeuwijk, F.C. Hillier, C.L. Bambra, R.M. van Rijn, and A. Burdorf, "The Contribution of Overweight, Obesity, and Lack of Physical Activity to Exit from Paid Employment: A Meta-Analysis," *Scandinavian Journal of Work, Environment & Health* 39, (2013); S.J. Robroek, T.I. van den Berg, J.F. Plat, and A. Burdorf, "The Role of Obesity and Lifestyle Behaviours in a Productive Workforce," *Occupational and Environmental Medicine* 68 (2011): 134–39.

26. Puhl et al., "Potential Policies and Laws"; Rhode, "The Injustice of Appearance," *Stanford Law Review* 61 (2009).

27. BMI (also known in parts of Europe as the Quételet index, after Adolphe Quételet (1796–1874), the Belgian statistician and social scientist who devised it) is calculated by dividing the weight of the body in kilograms by the height in meters (squared)—or dividing the weight of the body in pounds by the height in inches squared, then multiplying the result by 703.

28. See, for example, Center for Disease Control and Prevention (CDC), "BMI Percentile Calculator for Child and Teen (English Version)," https://nccd.cdc.gov/dnpabmi/Calculator.aspx; and "Adult BMI Calculator," www.cdc.gov/healthyweight/assessing/bmi/adult_BMI/english_bmi_calculator/bmi_calculator.html.

29. Obesity Foundation India, "Body Mass Index, BMI Calculator and Measuring Obesity in Children," Obesity Foundation India, http://obesityfoundationindia.com/bmi.htm.

30. See results https://nccd.cdc.gov/dnpabmi/Result.aspx?dob=1/1/2007&dom=1/2/2017&age=120&ht=100&wt=25&gender=1&method=1.

31. "Am I at Risk?" Joslin Diabetes Center: Asian American Diabetes Initiative, https://aadi.joslin.org/en/am-i-at-risk/asian-bmi-calculator.

32. T.N. Jayasinghe, V. Chiavaroli, D.J. Holland, W.S. Cutfield, and J.M. O'Sullivan, "The New Era of Treatment for Obesity and Metabolic Disorders: Evidence and Expectations for Gut Microbiome Transplantation," *Frontiers in Cellular and Infection Microbiology* 6 (2016): 1–11. P.J. Parekh, L.A. Balart, and D.A. Johnson, "The Influence of the Gut Microbiome on Obesity, Metabolic Syndrome and Gastrointestinal Disease," *Clinical and Translational Gastroenterology* 6 (2015): 1–12.

33. Government Office for Science (U.K.), *Tackling Obesities: Future Choices—Project Report,* Foresight Projects (London: Department of Innovation, Universities and Skills, 2007).

34. See, for example, Jean-Pierre Poulain, "Obesity and the Medicalization of Everyday Food Consumption," in *The Sociology of Food,* trans. Augusta Dörr, 81–111 (London: Bloomsbury, 2017).

35. Claude Fischler, "The 'Mcdonaldization' of Culture," in *Food: A Culinary History from Antiquity to the Present,* ed. J.-L. Flandrin and M. Montanari, trans. A. Sonnenfeld, 530–47 (New York: Columbia University Press, 1999); Claude Fischler, "Le Bon et le Saint, Évolution de la Sensibilité Alimentaire des Français," *Cahiers de l'OCHA,* (1993).

36. Claude Fischler, "Gastro-Nomie et Gastro-Anomie, Sagesse du Corps et Crise Bioculturelle de l'alimentation Moderne," *Communications* 31 (1979).

37. Nick Fiddes, "Declining Meat: Past, Present . . . And Future Imperfect?" in *Food, Health and Identity,* ed. Pat Caplan (London: Routledge, 1997), 255.

38. Carlo Petrini, *Slow Food: The Case for Taste* (New York: Columbia University Press, 2003), xiv.

39. J. Beaumont et al., *Report from the Policy Sub-Group to the Nutrition Task Force: Low Income Project Team* (Watford: Institute of Grocery Distribution, 1995).

40. See www.ers.usda.gov/data-products/food-access-research-atlas/go-to-the-atlas.

41. Accessible at www.fooddeserts.org.

42. Tim Lang and Martin Caraher, "Access to Healthy Foods Part II. Food Poverty and Shopping Deserts," *Health Education Journal* 57 (1998).

43. D.M. Cutler, E.L. Glaeser, and J.M. Shapiro, "Why Have Americans Become More Obese?" *The Journal of Economic Perspectives* 17, (2003): 93–118.

44. Ibid.

45. Ibid.

46. R.I. Weiss and J.A. Smith, "Legislative Approaches to the Obesity Epidemic," *Journal of Public Health Policy* 25 (2004): 379–390.

47. Guthman, *Weighing In,* 8–9.

48. Cutler, Glaeser, and Shapiro, "Why Have Americans Become More Obese?"

49. Poulain, *Sociology of Food,* 106.

50. Sarah Hinde and Jane Dixon, "Changing the Obesogenic Environment: Insights from a Cultural Economy of Car Reliance," *Transportation Research Part D* 10 (2005): 31–53.

51. Ibid.

52. See Guthman, *Weighing In.*

53. World Health Organization (WHO), "Obesity," www.who.int/topics/obesity/en.

54. *Report of the Commission on Ending Childhood Obesity* (Geneva: WHO, 2016), 2.

55. Poulain, *Sociology of Food,* 87–89.

56. Julier, "The Political Economy of Obesity," 484.

57. "Irresponsible, Careless Kids at Obesity Risk," *Times of India* August 22, 2013.

58. Susie Orbach, *Fat Is a Feminist Issue* (New York: Paddington Press, 1978), 18–19.

59. See http://ebonylifetv.com/programming/ebonylife-homegrown/reality/fattening-room.

60. Poulain, *Sociology of Food,* 103.

61. Emily Massara, "Que Gordita," in *Food and Culture: A Reader,* ed. C. Counihan and P. Van Esterik (New York: Routledge, 1997), 252.

62. Elisa J. Sobo, "The Sweetness of Fat: Health, Procreation, and Sociability in Rural Jamaica," in *Food and Culture: A Reader,* ed. C. Counihan and P. Van Esterik (New York: Routledge, 1997), 269.

63. Kelly D. Brownell, "Culture Matters in the Obesity Debate," *Los Angeles Times* September 21, 2007.

64. Poulain, *Sociology of Food,* 92.

65. François Xavier Lanthenas, *De l'influence de la Liberté sur la Santé, la Morale et le Bonheur* . . . (Paris: Impr. du Cercle social, 1792), 8.

66. Michel Foucault, *The Birth of the Clinic* (London: Routledge, 2003), 38.

67. Harvey A. Levenstein, "The Perils of Abundance: Food, Health, and Morality in American History," in *Food,* J.-L. Flandrin and M. Montanari, trans. A. Sonnenfeld (New York: Columbia University Press, 1999), 526.

68. Ibid., 527.

69. Ibid., 528.

70. See www.food.gov.uk/multimedia/pdfs/lidnssummary.pdf.

71. Warren Belasco, "Body and Soul," in *A Cultural History of Food in the Modern Age,* ed. Amy Bentley (London: Berg, 2012).

72. Ibid., 182.

CHAPTER 12 CONSUMPTION

1. Elijah Quashie and Elishama Udorok, "Chicken Connoisseur," YouTube channel, www.youtube.com/channel/UCZFmxd9L1btyXig-2Dp9XnA/featured. All of the Chicken Connoisseur's videos can be seen there.

2. From the "About" page at the Chicken Connoisseur's YouTube channel. Besides *mandem* and *creps,* other frequently used words in his videos are *peng* (extremely attractive and very well built) and *hench* (physically fit) both usually applied to people. For obvious reasons, when applied to food, *peng* retains its positive associations, while *hench* generally does not. For further advice, you can consult online sources such as the Urban Dictionary, www.urbandictionary.com.

3. U.K. television personality Gregg Wallace is a former greengrocer and long-term judge on the U.K. hit television cooking competition series *Masterchef,* where he offers the judgment of an "ordinary" (non-chef) but highly experienced consumer of "fine dining" restaurant food.

4. Simon Usborne, "The Chicken Connoisseur on His New Pengest Munch," *Guardian,* December 16, 2016.

5. Signe Hansen, "Society of the Appetite," *Food, Culture & Society* 11 (2008).

6. Emma-Jayne Abbots, "The Intimacies of Industry," *Food, Culture & Society* 18 (2015).

7. See, for example, the recipe for "Middle-School Tacos" on the cooking page of the *New York Times:* https://cooking.nytimes.com/recipes/1018758-middle-school-tacos.

8. Peter Naccarato and Kathleen LeBesco, *Culinary Capital* (London: Berg, 2012), 41.

9. Jean Anthelme Brillat-Savarin, *The Physiology of Taste,* trans. M.F.K. Fisher (San Francisco: North Point Press, 1986), 173–78.

10. Jean Anthelme Brillat-Savarin, *The Philosopher in the Kitchen,* trans. Anne Drayton (Harmondsworth, U.K.: Penguin, 1970), 155.

11. Ibid., 135.

12. Ibid., 157–58.

13. Ibid., 159.

14. Naccarato and LeBesco, *Culinary Capital,* 81. Kathleen LeBesco, *Revolting Bodies?* (Amherst: University of Massachusetts Press, 2004).

15. Thorstein Veblen, *The Theory of the Leisure Class* (New York: Macmillan, 1912), 68.

16. Ibid., 68–101 (chapter 4, "Conspicuous Consumption").

17. Ibid., 35–67 (chapter 3, "Conspicuous Leisure").

18. Ibid., 70.

19. Ibid., 73.

20. Ibid., 74.

21. Ibid., 75–76.

22. Alan Warde, *Consumption, Food and Taste* (London: Sage, 1997), 97.

23. Sidney Plotkin and Rick Tilman, *The Political Ideas of Thorstein Veblen* (New Haven, CT: Yale University Press, 2011), 145.

24. Ibid., 146.

25. Ian Burkitt, *Social Selves* (London: Sage, 2008), 3.

26. Plotkin and Tilman, *Political Ideas of Thorstein Veblen,* 6.

27. C.B. Macpherson, *The Political Theory of Possessive Individualism* (Oxford: Oxford University Press, 2011), 271–72.

28. Naccarato and LeBesco, *Culinary Capital,* 68.

29. Guy Debord, *La Société Du Spectacle* (Paris: Buchet/Chastel, 1967).

30. Guy Debord, *Society of the Spectacle* (Detroit: Black and Red, 1983), 202.

31. Anne-Marie Eyssartel, Bernard Rochette, and Jean Baudrillard, *Des Mondes Inventés: Les Parcs À Thème suivi de Biosphère II* (Paris: Editions de la Villette, 1992).

32. Will Self, "Guy Debord's the Society of the Spectacle," *Guardian* November 14, 2013.

33. Guy Debord, *Comments on the Society of the Spectacle,* trans. Malcolm Imrie (London: Verso, 1990).

34. Veblen, *Theory of the Leisure Class,* 70.

35. Naccarato and LeBesco, *Culinary Capital,* 32.

36. Ibid.

37. Ibid., 22. The authors focus on American consumers, but their analysis could apply equally to food advertising in other countries.

38. Ibid., 21.

39. Beth McCarthy-Miller, "Three Turkeys," *Modern Family* (excerpt online at http://abc.go.com/shows/modern-family/video/pl5520993/VDKA0_27rqrm8g, ABC, 2014).

40. Charlotte Brunsdon, "Feminism, Postfeminism, Martha, Martha, and Nigella," *Cinema Journal* 44 (Winter 2005): 114.

41. "Everyday Easy (Caramel Croissant Pudding)," YouTube video, 5:09, from *Nigella Express,* posted by "Fiona Super," February 3, 2010, www.youtube.com/watch?v=DJWdWUZNbow&list=PL244z0CgC2y49Kj9sTA_oEkSMjeP1qOWS.

42. Naccarato and LeBesco, *Culinary Capital,* 42.

43. Martha Rosler, *Semiotics of the Kitchen,* 1975. PERFORMANCELOGIA Performance Art Archive.

44. Electronic Arts Intermix, "Semiotics of the Kitchen, Martha Rosler," http://web.archive.org/web/20061006090705/http://www.eai.org/eai/tape.jsp?itemID=1545.

45. One of the many tributes to (or parodies of) Rosler's work is a stop-motion animation starring a Barbie doll. The discrepancies of scale between Barbie and her various implements, as well as the negligible space she has to manoeuver in, nicely emphasize the constraints raised by Rosler in her original work. See, "Semiotics of the Kitchen 2011 (Barbie Stop Motion)" YouTube video, 5:40, posted by "Thressa Willett," May 11, 2011, www.youtube.com/watch?v=Ca2RP5TQAxI, 2011.

46. Brunsdon, "Feminism, Postfeminism, Martha, Martha, and Nigella," 111.

47. Martha Rosler, "The Art of Cooking: A Dialogue between Julia Child and Craig Claiborne (Excerpt)." *e-flux journal* 65 SUPERCOMMUNITY, May-August 2015.

48. Elizabeth David, *French Provincial Cooking* (London: Michael Joseph, 1960).

49. Rosler, "Art of Cooking," 3.

50. Austin de Croze, *What to Eat and Drink in France* (London: Frederick Warne, 1931). The other eight arts were architecture, sculpture, painting, engraving, music and dancing, literature and poetry, the cinema, and fashion.

51. Rosler, "Art of Cooking," 4.

52. Gwen Hyman, "The Taste of Fame: Chefs, Diners, Celebrity, Class" *Gastronomica: The Journal of Food and Culture* 8 (2008): 43.

53. Ibid., 44.

54. See, for example, the original cover of Bourdain's, *Kitchen Confidential* (2000), and the tales in it and *Medium Raw* (2010), as well as the covers of White's cookbooks, such as *White Heat* (1990), and his biography, *The Devil in the Kitchen* (2007).

55. Matthew Wilson, "Groove Armada's Andy Cato, the DJ Who Became a Farmer in France," *Financial Times,* January 5, 2017.

56. Dave Simpson, "Vegan Black Metal Chef, Reggae Reggae Sauce, Headbanger's Kitchen: Why Do So Many Musicians Cook?" *Guardian,* July 23, 2015.

57. Hannah Ellis-Petersen, "Singer-Turned-Chef Kelis Pops up in London Restaurant," *Guardian,* July 4, 2016.

58. John Hind, "Kelis: I'm a Really Emotional Cook," *Guardian,* June 19, 2010.

59. Brian Manowitz, "VeganBlackMetalChef," YouTube channel, www.youtube.com/channel/UCp3iXxis9n_E_GfbE-_ksFw.

60. Ferran Adrià, Heston Blumenthal, Thomas Keller, and Harold McGee, "Statement on the 'New Cookery,'" *Guardian,* December 10, 2006, www.theguardian.com/uk/2006/dec/10/foodanddrink.obsfoodmonthly.

61. *Heston Blumenthal's—Kitchen Chemistry—Beef—Part 1 of 2* (Discovery Channel, 2010).

62. "175 Faces of Chemistry—Heston Blumenthal," YouTube video, 5:32, posted by "Royal Society of Chemistry," February 16, 2016, www.youtube.com/watch?v=6x4WmTpFEQo.

63. Nathan Myhrvold, Chris Young, and Maxime Bilet, *Modernist Cuisine: The Art and Science of Cooking,* 6 vols. (Bellevue, WA: The Cooking Lab LLC, 2011).

64. Gourmand International, "Gourmand Awards Winners 1995–2014," www.cookbookfair.com/index.php/gourmand-awards/winners-1995–2014-gg.

65. Modernist Cuisine, "Modernist Cuisine Overview [Archived]," http://modernistcuisine.com/books/modernist-cuisine.

66. Nicolas Chatenier, *Mémoires De Chefs* (Paris: Textuel, 2012).

67. Hyman, "Taste of Fame," 44.

68. Bill Buford, *Heat* (London: Jonathan Cape, 2006).

69. Priscilla Parkhurst Ferguson and Sharon Zukin, "The Careers of Chefs," in *Eating Culture,* ed. Ron Scapp and Brian Seitz, 92–111 (New York: State University of New York Press, 1998).

70. Veblen, *Theory of the Leisure Class,* 131.

Bibliography

Albala, K., ed. *Routledge International Handbook of Food Studies.* London and New York: Routledge, 2014.

Appadurai, A. "How to Make a National Cuisine: Cookbooks in Contemporary India." *Comparative Studies in Society and History* 30, (1998): 3–24.

Ashby, C., T. Gronberg, and S. Shaw-Miller, eds. *The Viennese Café and Fin-de-Siècle Culture* New York: Berghahn, 2013.

Barker, G. *The Agricultural Revolution in Prehistory* Oxford: Oxford University Press, 2006.

Basu, S. *Curry: The Story of Britain's Favourite Dish.* New Delhi: Harper Collins, 1999.

Bates, R. H. *Open-Economy Politics: The Political Economy of the World Coffee Trade.* Princeton, NJ: Princeton University Press, 1997.

Bell, D., and G. Valentine. *Consuming Geographies: We Are Where We Eat.* London: Routledge, 1997.

Bellwood, P. *First Farmers: The Origins of Agricultural Societies.* Oxford: Basil Blackwell, 2005.

Bickham, T. "Eating the Empire: Intersections of Food, Cookery and Imperialism in Eighteenth-Century Britain." *Past and Present* 198 (2008): 71–109.

Biltekoff, C. *Eating Right in America.* Durham, NC: Duke University Press, 2013.

Bottero, J. *The Oldest Cuisine in the World: Cooking in Mesopotamia.* Chicago: Chicago University Press, 2004.

Bourdieu, P. *Distinction: A Social Critique of the Judgement of Taste.* Translated by R. Nice. Cambridge, MA: Harvard University Press, 1984.

Brillat-Savarin, J. A. *The Philosopher in the Kitchen.* Harmondsworth, U.K.: Penguin, 1970.

———. *The Physiology of Taste.* Translated by M. F. K. Fisher. San Francisco: North Point Press, 1986.

Brown, J. "Ale House Licensing and State Formation in Early Modern England," in *Intoxication and Society: Problematic Pleasures of Drugs and Alcohol,* edited by J. Herring, C. Regan, D. Weinberg, and P. Withington, 110–32. London: Palgrave Macmillan, 2013.

Brown, L. R. *Full Planet, Empty Plates: The New Geopolitics of Food Scarcity.* New York: W. W. Norton, 2012.

Bynum, C. *Holy Feast and Holy Fast: The Religious Significance of Food to Medieval Women.* Berkeley: University of California Press, 1987.

Carney, J. A., and R. N. Rosomoff. *In the Shadow of Slavery: Africa's Botanical Legacy in the Atlantic World.* Berkeley: University of California Press, 2009.

Carolan, M. *The Sociology of Food and Agriculture.* 2nd ed. London and New York: Routledge and Earthscan, 2016.

Chehabi, H. "How Caviar Turned Out to Be Halal," *Gastronomica* 7 (2007): 17–23.

Clapp, J. *Food.* Cambridge: Polity Press, 2012.

Clarke, P. *The English Alehouse: A Social History.* London: Longman, 1983.

Collingham, L. *Curry: Tales of Cooks and Conquerors.* Oxford: Oxford University Press, 2006.

Cook, G. *Genetically Modified Language: The Discourse of Arguments for GM Crops and Food.* London: Routledge, 2004.

Cope Saguy, A. *What's Wrong with Fat?* New York: Oxford University Press, 2013.

Counihan, C., and P. van Esterik, eds. *Food and Culture: A Reader.* London and New York: Routledge, 2008.

Cowan, B. *The Social Life of Coffee: The Rise of the Emergence of the English Coffee-House.* New Haven, CT: Yale University Press, 2005.

Crosby, A. Jr. *The Columbian Exchange: Biological and Cultural Consequences of 1492.* Westport, CT: Greenwood, 1972.

———. *Ecological Imperialism: The Biological Expansion of Europe, 900–1900.* 2nd ed. Cambridge: Cambridge University Press, 2004.

Cullather, N. *The Hungry World: America's Cold War Battle Against Poverty in Asia.* Cambridge, MA: Harvard University Press, 2010.

Cusack, I. "African Cuisines: Recipes for Nation-Building?" *Journal of African Cultural Studies* 13 (2000): 207–25.

Daviron, B., and S. Ponte. *The Coffee Paradox: Global Markets, Commodity Trade and the Elusive Promise of Development.* London: Zed, 2005.

Desmarais, A. A. *La Vía Campesina: Globalization and the Power of Peasants.* London: Pluto Press, 2007.

Dickie, J. *Delizia! The Epic History of Italians and Their Food.* New York: Free Press, 2008.

Douglas, M. "Deciphering a Meal," *Daedalus* 101 (1972): 61–81.

———. *Purity and Danger.* London: Routledge, 2002.

Earle, R. *The Body of the Conquistador: Food, Race and the Colonial Experience of Spanish America, 1492–1700.* Cambridge: Cambridge University Press, 2014.

Elias, N. *The Civilizing Process: The History of Manners and State Formation and Civilization*. Oxford: Blackwell, 1994.

Ellis, M. *The Coffee-House: A Cultural History*. London: Weidenfeld & Nicolson, 2004.

Elmore, B. *Citizen Coke: The Making of Coca-Cola Capitalism*. New York: Norton, 2015.

Ferguson, P. *Accounting for Taste: The Triumph of French Cuisine*. Chicago: Chicago University Press, 2004.

Fernández-Armesto, F. *Near a Thousand Tables: A History of Food*. New York: Free Press, 2002.

Fischer, J. *The Halal Frontier: Muslim Consumers in a Globalized Market*. Edited by Laurel Kendall. New York: Palgrave Macmillan, 2011.

Fischler, C. "The 'Mcdonaldization' of Culture," in *Food: A Culinary History from Antiquity to the Present,* edited by J.-L. Flandrin and M. Montanari, translated by A. Sonnenfeld, 530–47. New York: Columbia University Press, 1999.

Foster, N., and L. S. Cordell. *Chiles to Chocolate: The Foods the Americas Gave the World*. Tucson: University of Arizona Press, 1992.

Freedmann, H. "From Colonialism to Green Capitalism: Social Movements and the Emergence of Food Regimes," in *New Directions in the Sociology of International Development: Research in Rural Sociology and Development,* edited by F. H. Buttel and P. D. McMichael, 227–64. Amsterdam: Elsevier, 2005.

Friedmann, H., and P. McMichael, "Agriculture and the State System: The Rise and Fall of National Agricultures, 1870 to the Present." *Sociologia Ruralis* 29 (1989): 92–117.

Gabaccia, D. R. "Colonial Creoles: The Formation of Tastes in Early America," in *Taste Cultures: Experiencing Food and Drink,* edited by C. Korsmeyer, 79–87. London: Bloomsbury, 2005.

———. *We Are What We Eat: Ethnic Food and the Making of Americans* Cambridge, MA: Harvard University Press, 1998.

Glaeser, E.-L., and J. M. Shapiro. "Why Have Americans Become More Obese?" *The Journal of Economic Perspectives* 17 (2003): 93–118 .

Goody, J. *Cooking, Cuisine and Class: A Study in Comparative Sociology*. Cambridge: Cambridge University Press, 1982.

Grimm, V. E. *From Feasting to Fasting: The Evolution of a Sin*. London: Routledge, 1996.

Guthman, J. *Weighing In: Obesity, Food Justice, and the Limits of Capitalism*. Berkeley: University of California Press, 2011.

Gvion, L. *Beyond Hummus and Falafel: Social and Political Aspects of Palestinian Food in Israel*. Translated by David Wesley and Elana Wesley. Berkeley: University of California Press, 2012.

Hailwood, M. *Alehouses and Good Fellowship in Early Modern England*. Woodbridge: Boydell, 2014.

Haine, W. S. *The World of the Paris Café: Sociability among the French Working Class, 1789–1914*. Baltimore, MD: Johns Hopkins University Press, 1996.

Harris, M. *Good to Eat*. London: Allen & Unwin, 1986.

Hattox, R. *Coffee and Coffeehouses: The Origins of a Social Beverage in the Medieval Near East.* Seattle: University of Washington Press, 1985.

Haydon, P. *The English Pub: A History.* London: Robert Hale, 1994.

Herring, R. J. *The Oxford Handbook of Food, Politics, and Society.* Oxford: Oxford University Press, 2015.

Ichijo, A., and R. Ranta. *Food, National Identity and Nationalism: From Everyday to Global Politics.* Basingstoke, U.K.: Palgrave Macmillan, 2016.

Jacobs Solomonson, L. *Gin: A Global History.* London: Reaktion, 2012.

Johnston, J., and S. Baumann. *Foodies: Democracy and Distinction in the Gourmet Foodscapes.* London: Routledge 2010.

Ketcham Wheaton, B. *Savoring the Past: The French Kitchen and Table from 1300 to 1789.* London: Chatto and Windus,1983.

Kümin, B. *Drinking Matters: Public Houses and Social Exchange in Early Modern Central Europe.* London: Palgrave Macmillan, 2007.

———. "Eating Out before the Restaurant: Dining Cultures in Early-Modern Inns," in *Eating Out in Europe: Picnics, Gourmet Dining, and Snacks Since the Late Eighteenth Century,* edited by M. Jacobs and P. Scholliers, 71–87. Oxford: Berg, 2003.

Laudan, R. *Cuisine and Empire: Cooking in World History.* Berkeley: University of California Press, 2013.

Levenstein, H. A. "The Perils of Abundance: Food, Health, and Morality in American History," in *Food,* edited by J.-L. Flandrin and M. Montanari, translated by A. Sonnenfeld, 516–26. New York: Columbia University Press, 1999.

Levi, J. "Disgust," in *The Sage Encyclopedia of Food Issues,* edited by Ken Albala, 367–70. Thousand Oaks, CA: Sage, 2015.

Lévi-Strauss, C. *The Raw and the Cooked.* Harmondsworth, U.K.: Penguin, 1986.

Magnan, A. "Food Regimes," in *The Oxford Handbook of Food History,* edited by J. M. Pilcher, 370–88. Oxford: Oxford University Press, 2012.

Matthee, R. "Alcohol in the Islamic Middle East: Ambivalence and Ambiguity," in *Cultures of Intoxication,* edited by P. Withington and A. McShane, 100–125. Oxford: Oxford University Press, 2014.

Mazumdar, S. "The Impact of New World Crops on the Diet and Economy of China and India, 1600–1900," in *Food in Global History,* edited by R. Grew, 58–78. Boulder, CO: Westview, 1999.

McGirr, L. *The War on Alcohol: Prohibition and the Rise of the American State.* New York: Norton, 2016.

McGovern, P. *Ancient Wine: The Search for the Origins of Viniculture.* Princeton, NJ: Princeton University Press, 2003.

McMichael, P. "A Food Regime Analysis of the 'World Food Crisis'". *Agriculture and Human Values* 26 (2009): 281–95.

McNeill, W. H. "How the Potato Changed the World's History." *Social Research* 66 (1999): 67–83.

McShane, A. "Material Culture and 'Political Drinking' in Seventeenth-Century England." *Past and Present,* supplement 9 (2014): 247–76.

Melville, E. G. K. *A Plague of Sheep: The Environmental Consequences of the Conquest of Mexico.* Berkeley: University of California Press, 1997.

Mennell, S. *All Manners of Food: Eating and Taste in England and France from the Middle Ages to the Present.* Urbana: University of Illinois Press, 1996.

Mennell, S., A. Murcott, and A.H. van Otterloo. *The Sociology of Food: Eating, Diet and Culture.* London: Sage, 1992.

Mintz, S.W. *Sweetness and Power: The Place of Sugar in Modern History.* London: Penguin, 1985.

Murcott, A. "Cooking and the Cooked: A Note on the Domestic Preparation of Meals," in *The Sociology of Food and Eating,* edited by A. Murcott, 178–93. Aldershot, U.K.: Gower, 1983.

———. "Family Meals: A Thing of the Past?" in *Food, Health, and Identity,* edited by P. Caplan, 32–49. London: Routledge, 1997.

Nabhan, G.P. *Cumin, Camels and Caravans: A Spice Odyssey.* Berkeley: University of California Press, 2014.

Naccarato, P., and K. LeBesco. *Culinary Capital.* London: Berg, 2012.

Nicholls, J. *The Politics of Alcohol: A History of the Drink Question in England.* Manchester, U.K.: Manchester University Press, 2011.

Norton, M. *Sacred Gifts, Profane Pleasures: A History of Tobacco and Chocolate in the Atlantic World* Ithaca: Cornell University Press, 2008.

———. "Tasting Empire: Chocolate and the European Internalization of Mesoamerican Aesthetics." *American Historical Review* 111 (2006): 660–91.

Ollivry, F. *Les secrets d'Alep: Une grande ville arabe revelee par sa cuisine.* Paris: Sinbad/Actes Sud, 2006.

Orbach, S. *Fat Is a Feminist Issue.* New York: Paddington Press, 1978.

Panayi, P. *Spicing Up Britain: The Multicultural History of British Food.* London: Reaktion, 2008.

Parker Talwar, J. *Fast Food, Fast Track: Immigrants, Big Business, and the American Dream.* Boulder, CO: Westview Press, 2004.

Parkhurst Ferguson, P., and S. Zukin. "The Careers of Chefs" in *Eating Culture,* edited by R. Scapp and B. Seitz, 92–111. New York: State University of New York Press, 1998.

Petrini, C. *Slow Food: The Case for Taste.* New York: Columbia University Press, 2003.

Phillips, R. *Alcohol: A History.* Chapel Hill: University of North Carolina Press, 2014.

Pilcher, J.M. *Food in World History.* London: Routledge, 2006.

Poulain, J.-P. *The Sociology of Food: Eating and the Place of Food in Society.* London: Bloomsbury, 2017.

Reader, J. *The Untold History of the Potato.* London: Vintage, 2008.

Rees, J. *Refrigeration Nation: A History of Ice, Appliances, and Enterprise in America.* Baltimore, MD: Johns Hopkins University Press, 2013.

Rich, R. *Bourgeois Consumption: Food, Space, and Identity in London and Paris, 1850–1914.* Manchester, U.K.: Manchester University Press, 2011.

Rozin, P., J. Haidt, C. McCauley, and S. Imada. "Disgust: Preadaptation and the Cultural Evolution of a Food-Based Emotion," in *Food Preferences and Taste,* edited by H. Macbeth, 65–82. Oxford: Berghahn, 1997.

Rubel, W. *Bread: A Global History.* London: Reaktion, 2012.

Salaman, R.N. *The History and Social Influence of the Potato*. 2nd ed. Cambridge: Cambridge University Press, 2010.

Salinger, S. *Taverns and Drinking in Early America*. Baltimore, MD: Johns Hopkins University Press, 2002.

Schwarz Cohan, R. *More Work for Mother: The Ironies of Household Technology from the Open Hearth to the Microwave*. London: Free Association, 1989.

Sen, A. *Poverty and Famines: An Essay on Entitlement and Deprivation*. Oxford: Oxford University Press, 1990.

Sharma, J. "Food and Empire," in *The Oxford Handbook of Food History*, edited by J.M. Pilcher, 241–57. Oxford: Oxford University Press, 2012.

Simmel, G. "Sociology of the Meal," in *Simmel on Culture: Selected Writings*, edited by M. Featherstone, 130–35. London: Sage, 1997.

Simoons, F.J. *Eat Not This Flesh*. Madison: University of Wisconsin Press, 1961.

Smith, R.E.F., and D. Christian, *Bread and Salt: A Social and Economic History of Food and Drink in Russia*. Cambridge: Cambridge University Press, 1984.

Sournia, J.-C. *A History of Alcoholism*. Oxford: Basil Blackwell, 1990.

Spang, R. *The Invention of the Restaurant: Paris and Modern Gastronomic Culture*. Cambridge, MA: Harvard University Press, 2000.

Tansey, G., and T. Worsley. *The Food System: A Guide*. London: Earthscan, 1996.

Toussaint-Samat, M. *A History of Food*. Translated by A. Bell Oxford: Wiley-Blackwell, 2009.

Trauger, A., ed., *Food Sovereignty in International Context: Discourse, Politics and Practice of Place*. London: Routledge, 2015.

Walvin, J. *Fruits of Empire: Exotic Produce and British Taste, 1660–1800*. Houndsmills and London: Macmillan Press, 1997.

Warde, A. *Consumption, Food and Taste: Culinary Antinomies and Commodity Culture*. London: Sage, 1997.

Warner, J. *Craze: Gin and Debauchery in an Age of Reason*. London: Profile, 2003.

Watts, S. "Food and the Annales School," in *The Oxford Handbook of Food History,* edited by J.M. Pilcher, 3–22. Oxford: Oxford University Press, 2012.

Weis, T. *The Global Food Economy: The Battle for the Future of Farming*. London: Zed, 2007.

Wild, A. *Black Gold: The Dark History of Coffee*. London: Harper Perennial, 2005.

Wilk, R. *Home Cooking in the Global Village: Caribbean Food from the Buccaneers to Ecotourists*. Oxford: Berg, 2006.

Wrangham, R. *Catching Fire: How Cooking Made Us Human*. London: Profile, 2009.

Zubaida, S. "Drink, Meals and Social Boundaries," in *Food Consumption in Global Perspective: Essays in the Anthropology of Food in Honour of*

Jack Goody, edited by J. Klein and A. Murcott, 209–23. London: Palgrave, 2014.

———. "The Idea of 'Indian Food': Between the Colonial and the Global." *Food and History* 7 (2009): 191–210.

———. "National, Communal and Global Dimensions in Middle Eastern Food Cultures," in *A Taste of Thyme: Culinary Cultures of the Middle East*, edited by S. Zubaida and R. Tapper, 33–45. London: Tauris, 2000.

Index